Minerals from the Marine Environment

Sir Peter Kent, D.Sc., F.R.S.

former Chairman, Natural Environment Research Council

With contributions from Dr. N. C. Fleming,
Institute of Oceanographical Sciences

Edward Arnold

©P. E. Kent 1980

First published 1980 by
Edward Arnold (Publishers) Ltd
41 Bedford Square, London WC1B 3DQ

British Library Cataloguing in Publication Data

Kent, Percy Edward
    Minerals from the marine environment. — (Resource
    and environmental sciences series).
    1. Mineral industries 2. Mineral resources in
    submerged lands 3. Marine mineral resources
    I. Title   II. Series
    333.8   HD9506.A2

    ISBN 0 7131 2813 5

Typeset by The Castlefield Press of Northampton

Printed in Great Britain
by Spottiswoode Ballantyne Limited, Colchester and London

Preface

The word 'environment' has come to mean widely different things to different people. At one extreme it represents the current milieu, which has to be preserved at all costs, in a condition as nearly unchanged as possible. To others it is the natural order of things, the combination of external circumstances, which is as often adverse as favourable for such human affairs as food production and maintenance of life, and which may have to be fought against in terms of pests, disease and natural disaster.

Both concepts are valid and important, but there is a third category of environmental factors — those exercised by the long history of development of the continents and oceans; the changing physical and chemical conditions which through geological time to the present have left their influence on the distribution of elements both useful and useless, and which have direct relevance to their employment for human welfare. It is this aspect of environmental control which is particularly emphasized in the following pages.

Pollution of the environment is an important aspect of a majority of human activities, from food manufacturing to burning garden rubbish, and in this range mineral extraction is no exception. In a world of progressively increasing population there must be a progressive *potential* increase in pollution, and it is necessary now to develop an increasingly sophisticated understanding of natural phenomena to minimize — and if possible reverse — the disturbing effects of man's activities.

On land the conflicting requirements of development and natural environment are well documented and understood, although solutions may not be obvious. The seas, covering three quarters of our globe, contain enormous undeveloped resources for the future, but at the same time they are the ultimate and final sink for the wastes of the lands. Some of the special problems of the pollution of the seas are mentioned in this book, although the utilization of the seas and their minerals is still at an early stage. Mankind must have energy and materials for survival, and the seas will make their very large contribution — provided that the resultant problems can be overcome and the resources provided without too great an adverse balance or disadvantage in disturbing the natural order.

The draft of this book benefitted markedly from advice from Dr Nicholas Fleming of the Institute of Oceanographic Sciences, who provided useful criticism and contributed particularly to Chapter 2.

Warm thanks are due to The British Petroleum Company for Figs 7.3, 7.5, 7.8, 7.10 and 7.12 and for basic data for several diagrams; to the Institute of Oceanographical Sciences for Figs. 2.1, 2.5, 2.6, 2.7 and 3.6; to the Westminster Dredging Company for Fig. 3.1 and to the Institute of Geological Sciences for Fig. 4.2.

Miss Muriel Bransdon converted my complex manuscript into legible text and Mr Alan Miles drew the diagrams.

1980   P.E.K.

Contents

1 Introduction

With good reason, concern about the long-term availability of mineral resources has become accentuated as their exploitation on land has required recovery from less and less rich sources, leading to concern over exponential growth expressed in the studies of the group of economic forecasters who comprise the Club of Rome and illustrated in *The Limits to Growth* (Meadows, *et. al.*, 1972). Under these pressures, the importance of the seas to supplement the reserves of the land has increased and is more and more the subject of research and development of new technologies, with even the deep oceans as a practical target for metal and perhaps hydrocarbon reserves.

The petroleum industry provides the most familiar example of development of offshore resources, which followed the initial extension of coastal oilfields into the shallow waters of the Gulf of Mexico and the Caspian Sea in the early years of this century. Technology has developed to the point where production is now established beyond the 100-fathom line, the traditional edge of the continental shelf. It is becoming technically and economically practical to find and produce oil and gas in water depths considerably below this; drilling rigs are now available for controlled drilling and testing in 2000m of water, with remote controlled sea-floor equipment for production now under test.

Although public attention is focussed on the practical problems of petroleum extraction on the continental shelves, and on the political problems of metal recovery from manganese nodules in the deep oceans, (see Chapter 5) it must be emphasized that a far larger tonnage of the humble bulk minerals — sand and gravel, calcium carbonate sand — is already being extracted from the shallow seas. To some extent these are renewable reserves, although like any other aspect of the natural environment the renewal mechanism can be locally overloaded, but for these and also for the dissolved constituents of ocean water the total quantity in existence — the ultimate reserve, not necessarily economically recoverable — is of astronomical size.

Each group of minerals has its separate range of problems, related to the environmental conditions of formation and, even more, to the environmental constraints on recovery. It is the purpose of this small book to illustrate the problems and their present and possible future solutions for the effective and non-injurious development of the world's largest mineral store.

The literature on the subject is voluminous, relating not only to the occurrence of offshore mineral wealth but to the ocean environment and the very wide range of economic and engineering problems which arise in developing its potential. Only a scatter of references is given, but these should enable the reader to follow up problems in the wider literature.

2 The Marine Environment

As compared with the land, the sea is a hostile environment for mineral exploitation; it is unstable, unpredictable and has the potential for wrecking most of man's structural efforts. It has some advantages such as (mainly) unobstructed passage for transport, the vast bulk available for containing even trace resources, and in the ebb and flow which constantly renews the water in any given area, and which dilutes the polluting material discharged into the sea in such large quantities by our civilization.

Whether the seeking and finding of minerals can take advantage of marine phenomena or must be carried out in spite of them, the fullest possible understanding of the physical, chemical and organic processes of the seas and oceans is necessary to optimize the results and to minimize potential environmental damage. Some of the factors involved are outlined in the following paragraphs, but it must be emphasized that the body of available data on these subjects is enormous.

In the western world there are numerous academic and commercial establishments specializing in acquisition of marine data (obtained, for example, from ships such as the one shown in Fig. 2.l), processing and storage in different fields — corrosion, marine meteorology, submarine geological structures, currents, bathymetry, gravity, and so on. In most of the developed countries there are National Oceanographic Data Centres linked to the Oceanographic Data Centre of the World Data Centre. Within countries the National Oceanographic Data Centres can contact the specialist agencies when they do not hold the required data themselves. In Britain the central point for marine information is the Marine Information and Advisory Service (MIAS), of the Natural Environment Research Council (NERC) Institute of Oceanographic Sciences. MIAS has access to its own data bank, and to all the expertise and data of the Institute of Oceanographic Sciences, as well as close contact with numerous other institutions and companies. It is necessary to consult specialist services like these if the environmental problems of the sea are to be fully understood.

The Origin of the Oceans

In the earlier stages of cooling from an incandescent body the surface of the world was far too hot to contain seas or lakes on its surface; any

Fig. 2.1 The oceanic research ship, 'Discovery'. (Photograph by courtesy of the National Institute of Oceanography.)

rain which fell flashed off immediately into steam. As the cooling process proceeded boiling water drained from the higher ground into the first seas and oceans, with a degree of chemical erosion far beyond any known now. Among the sediments resulting from this process are massive haematitic ironstones found in early Precambrian rocks. (See Glossary I for geological periods and time scale.) As the seas cooled and expanded the earliest recognizable life forms are algae, which presumably lived by photosynthesis, and simple objects which may be bacteria and could have been anaerobic. These are recognized in rocks a thousand million years older than those containing the animal precursors of today's marine fauna.

It has been assumed that the transport of soluble salts from land to sea via the rivers was essentially a one-way movement; that, except for the occasional burial of salts in sediments due to local evaporating dish traps and/or such occasional cataclysms as water spouts, land derived salts accumulated progressively in the oceans. Estimates of the age of the earth were made by relating the modern salinity to the rate of solubles transported by the world's rivers. However, it is now known that large scale reaction of ocean water with newly formed volcanic crust takes place on the mid-ocean ridges, and that a great bulk of salt-saturated sediments is taken into the continental masses along subduction planes so that we have a mechanism for returning oceanic salts to the crust. It is no longer necessary to assume that the oceans were progressively fresher the further back in time we look, and it becomes realistic to accept that the marine organisms of the last 900 million years — allied, although distantly, to those of today — lived under similar envirnonmental conditions of temperature and salinity.

The relative distribution of continents and oceans has not remained fixed in space. Over hundreds of millions of years continents have split and rejoined, some oceans have grown and others have closed. This process is still going on. Thus the Atlantic and Indian Oceans are widening at the rate of a few centimetres a year, while large parts of the Pacific oceanic coast are being overridden by the surrounding marginal plates. This concept, supported in considerable detail by the major advances in knowledge of ocean structure, is relevant to the distribution of minerals past and present, on land and at sea, partly because of resulting fundamental changes in the marine circulation pattern and partly because of the chemical processes accompanying the rifting of some oceans and the telescoping of others.

The tensional forces are concentrated in the mid-ocean ridges, the world-encircling submerged mountain belts of rifted basalts in which new igneous rock is added centrally as the plates on either side move apart (Fig. 2.2). So far they are not the sites of mineral recovery, but it has recently been found that they are marked by emergence of hot springs which deposit metalliferous muds, a possible potential for the future.

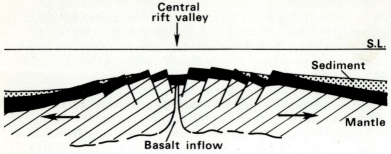

Fig. 2.2 Diagram showing the structure of a mid-ocean ridge. Metalliferous deposits are found in connection with hot springs in fault lines.

The constant access of new crust along mid-ocean ridges has to be compensated by disappearance of the equivalent area elsewhere on the world's surface. This takes place in 'subduction zones', usually associated with island arcs, chains of volcanoes, and frequent earthquakes. Under western South America and parts of the north-western United States the oceanic crust slopes down directly under the continent, and it is widely believed that extensive metalliferous deposits found in the Western Cordillera of the Americas represent concentrates from the ocean crust mobilized in this process (see Fig. 2.3). For example, manganese nodules may be part of the sediment subducted into the crust of the earth.

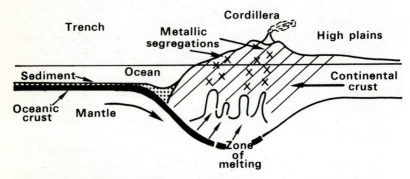

Fig. 2.3 Structure of a compressional (subduction) Pacific type of continental margin. Economic mineral deposits on land can be explained as ocean floor segregations which have been re-mobilized.

In the more general case the oceanic crust does not descend directly under a continent, but beneath an island arc some hundreds to a thousand kilometres seaward from the continent. As the oceanic crust descends to depths of several hundred kilometres the friction, breaking-up of the slab, and melting of sediments produce earthquakes, while the lighter molten materials rise up to form a chain of volcanic islands in the marginal sea. Such features are found around the entire western, southern and northern borders of the Pacific, the Scotia Arc, Caribbean, and to a certain extent in the Mediterranean Calabrian and Hellenic arcs.

The Continental Edges

One result of the mechanism of oceanic opening and closing is that none of the continental margins is old, relative to the time marine life has proliferated. But the modern pattern was in fact blocked out some 280 million years ago; the establishment of true oceanic depths in the Atlantic and Indian oceans dates back 70–100 million years. These periods have provided ample time for the processes of continental erosion, river and coastal transport and sedimentation to build up great thicknesses of sedimentary rocks which form the continental shelves, continental slopes and rises around the modern oceans (see Fig 2.4).

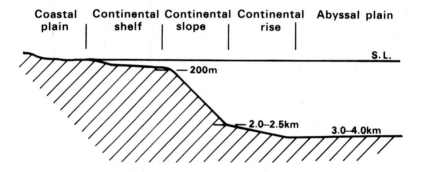

Fig. 2.4 The structure of a typical continental margin. The depths of changes in gradient vary appreciably in different regions.

The earlier rocks laid down around the newly formed oceans were deposited in rift valleys, some of them subaerial and containing porous rocks capable of acting as reservoirs, some the locus of salt and other evaporites, others again filled with thick fine grained marine muds rich in organic material which are potential source rocks for hydrocarbons. Up to 7000m of sediment were accumulated in this phase. These rifted basins were subsequently buried by a wedge of sediments deposited on the edge of what was by then the open ocean, sediments which make up the surface of the continental shelves, the bulk of the continental slopes and the continental rises, aggregating 10 000 m or more in thickness. The economic potential of these rocks, strongly dependent on environmental factors, is discussed in later chapters.

Water Movement in Seas and Oceans

Any activity at sea, whether transport or exploitation of resources, contrasts with operations on land in that beyond the shallows there is no fixity of foundations. Mineral recovery has to be carried out from a movable and moving base.

The complexity of water movement is still being elucidated. Tides in the open ocean are quite small in amplitude, and are governed by the combined gravitational effects of the sun and the moon. On the continental shelf and in enclosed seas and estuaries the effect of the tide is greatly magnified, and the rise and fall of the water as well as the tidal currents of several kilometres per hour can be a severe problem to ship operations and to construction work, problems accentuated since the time of the high and low tide precesses round the clock with the moon's gravitational effect. Figure 2.5 shows an encapsulated recording pressure gauge for measuring tidal water movements.

Fig. 2.5 A recording tide gauge. This instrument is placed on the sea floor and records tides by pressure measurements. (Photograph by courtesy of the Institute of Oceanographic Sciences.)

Waves are perhaps the most obvious and universal expression of the destructive energy of the sea — they can break a harbour wall, throw a ship onto rocks, fracture the legs of a drilling rig and erode away a beach or cliff. Waves are generated by the force of the wind blowing over the sea surface, and the height, energy, and length of the wave from crest to crest all increase with the strength and duration of the wind, and the greater distance over which the wind can act upon the sea surface. This last factor is known as the 'fetch'. After the wind has stopped blowing, or if the waves travel outside the wind area, the remaining waves are called 'swell'. It follows that at any particular place and time there may be residual waves from a nearby wind which has been blowing recently, a swell coming in from a distant storm hundreds or thousands of kilometres away, and fresh waves building up rapidly from a strong wind which started a few hours ago. Each of these wave patterns may be travelling in a different direction. It is the interaction of several wave trains like this which produces the typical confused and choppy sea. Since the maximum wave size from each direction depends on the fetch in that direction, any location on the continental shelf is likely to be exposed to quite different wave patterns from each direction. In addition, the larger waves are altered by interaction with the sea bottom, and with headlands and islands.

These factors are only a few of those which determine the combination of high and low waves of numerous periods passing a point, known as the wave spectrum. For different purposes an engineer may wish to know different things about the spectrum, or the frequency of recurrence of high waves, or calm periods, or the average wave height for a year, or the highest wave which could occur in 100 years, or the percentage occurrence of waves of a certain period which may set a ship or oil rig oscillating.

To describe, analyse, and predict the wave spectrum and to forecast the wave climate at many places clearly requires an enormous volume of measurements, and a lot of statistical analysis. To help co-ordinate the access to such data around the world MIAS operates the Responsible National Oceanographic Data Centre for Waves on a world-wide basis. The sophistication of the measuring devices is indicated by the structure of a data buoy located off Western Scotland (Fig. 2.6).

Additionally, surges caused by atmospheric effects may be of high importance in causing widespread coastal flooding or (in the case of negative surges) the danger of grounding of large vessels. Superimposed on these it has been found that very long period oceanic swells persist into shallow seas. The ultimate disaster may be caused by earthquake-generated waves known as 'tsunamis'. Factors such as these have to be taken into account in economic development at sea, whether in the design and maintenance of fixed and floating structures or in the transport of the products.

Aerial

Wind speed sensor
Wind direction sensor

Air pressure pitot head

Visibility meter

Rain gauge

Emergency
batteries

Gas
cylinder

Ballast tank

Power distribution
and miscellaneous
equipment

Main batteries

Mooring tension sensor

Acoustic current
meter spars

Thermo-mechanical
generator

Air temperature and
humidity sensors

Digital compass

Ballast tank

Cable well

Data handling
equipment

Sensor auxiliary
equipment

Roll and pitch sensor

Surface water
temperature sensor

Multicore cables to
underwater sensor packages

Heave sensor

Fig. 2.6 A, Photograph and **B,** structure of an Oceanographic Data Buoy. Several of these units have been sited off northern Scotland. (Photograph and diagram by courtesy of the Institute of Oceanographic Sciences.)

Currents at sea may be tidal, seasonal, or related to world circulation controls. It is now known that such surface currents as the Gulf Stream have great meanders and swirls, and that the deep circulation pattern in the Atlantic Ocean, for example, is largely different from that on the surface, with tidal oscillations and deep eddies superimposed on a general circulation of ocean water between the equator and the poles. A neutrally buoyant float at depth in the ocean will follow an extremely complex route of swings and spirals over an area of hundreds of miles within a few months, and equipment between surface and the ocean bottom is subjected to different stresses at different depths, stresses which are additionally subject to continual change.

The Sea Floor

There is no uniformity about the sea bottom, and its variations are critical for many operations. At one extreme it may be too hard and featureless for anchors to find any lodgement, as off parts of southern Australia, or it may be a uniform limestone surface as in the southern Persian Gulf. At the other extreme muddy water may grade down into deep watery mud, with no surface on which a diver can stand, and in which piles have to be driven scores of feet to obtain any security, as was found in parts of the Northern North Sea. The continental shelves are largely occupied by sands and muds or by tropical coral reefs; the continental slopes by silts, sands and muds; the deep ocean by a range of fine grained oozes; all products of erosion of the continents and transport of the debris either as particles or in solution.

Instability of fine muds presents one set of problems, either by sliding on natural slopes or by inability to stand on the sides of a cut channel, but coarser material may also be liable to movement. In the Gulf of Mexico collapse of oil-drilling rigs is recorded as being caused by slipping of unstable muds, and areas off the east coast of the U.S.A. have been excluded from petroleum licencing on the grounds of potential instability.

Moving sand waves are now a recognized phenomenon in the shallow seas; these may be several tens of metres high and hundreds of kilometres long; they move slowly under the influence of currents and storms and are liable to cause problems with buried sea floor pipelines, which may be first exhumed and then left unsupported. Presumably turbidity currents (sediment-loaded currents flowing down-slope under gravity) of the kind which have cut telegraph cables on the continental slope of north-eastern North America will also be a hazard to pipelines when these have to be established in areas subject to this form of high-speed sea floor erosion. The oil industry is also sharply conscious of the future problem in high latitudes of burying oil and gas lines sufficiently deeply to be out of reach of iceberg scour. Scour marks are common on the edge of the NW European shelf in 200 m water depth, illustrating the extent of the problem.

Establishment of foundations or the mining of the sea floor inevitably disturbs the dynamic equilibrium of the bottom water and sediments.

Local current scour is consequently a serious concern in designing fixed structures, which must be protected from resulting instability either by piling or by protective skirts. Similarly, dredging operations have to be planned with an adequate knowledge of the movement of bottom sediments, both as an aid to efficiency and for disposal of waste.

Sea Floor Surveys

Knowledge of the depth, form and nature of the sea floor is essential to navigation and critical for mineral development offshore, and survey methods of increasing sophistication have been developed to provide the necessary information.

The early methods of depth measurement using a lead line have been replaced by various forms of echo-sounding, asdic measuring depths continuously beneath a ship's keel, and oblique asdic, with its development in the 'Gloria' device (see Fig. 2.7), surveying a swathe some kilometres wide on each side of the survey ship and mapping features which might be missed by spot depth measurements or linear traverses. The broad picture thus produced requires adjustment for scale variation but provides valuable detail for mapping obstacles or the geology of the sea floor.

The nature of the floor — soft or hard, mud, sand, gravel or rock — was originally mapped by sampling with a piece of tallow in the sounding lead; there is now a range of samplers, grabs, gravity coring devices, and sea floor drills which can obtain more or less undisturbed samples of the sea floor sediments. The latest device is a hydraulic piston corer which can obtain completely undisturbed samples of sediments in deep water to tens or hundreds of metres below the sea floor. This is important whether the sea floor is itself mineral bearing (for phosphates, manganese nodules etc) or whether the critical question is stability for offshore structures. Shallow reflection surveys, such as the sparker method in which echo sounding penetrates the shallower sediments, can tie together such spot samples and demonstrate sedimentary structure in rapidly varying lithologies, and deep seismic surveys penetrate the sediments to depths of five kilometres or more in the search for the hydrocarbon riches of the continental shelves and slopes.

Most seismic reflection operations have hitherto been carried out in fairly shallow water (depths up to say 300 m) and environmental constraints arise in extending the method to great depth. Thus at present both energy source (usually an explosion) and recording seismometers are towed behind the operating ship near the sea surface, and in consequence are increasingly distant from the phenomena under study as the water depth increases. Sea water is not homogeneous in its physical character, with energy transmission in particular strongly affected by temperature differences of different layers. Accurate mapping of the structure of the lower continental slope and rise may depend on methods of recording data close to the sea floor.

Fig. 2.7 The recovery of 'Gloria' Mark I, a geological long-ranged asdic, after a side scan sea floor survey. (Photograph by courtesy of the Institute of Oceanographic Sciences.)

3 Bulk and Non-metallic Minerals from the Sea

Although the activities of the oil industry are more glamorous and the attempts to recover metals from the ocean floor excite the imagination, it must be remembered that enormous tonnages of the humdrum but essential bulk minerals, sand, gravel and limestone, are already recovered from the sea. Additionally, phosphates and diamonds provide important potential reserves, also recoverable by conventional dredging operations and sharing many of the same problems of water movement and recovery efficiency. One non-metallic mineral, barytes, is recovered by mining of the sea bed, and sulphur deposits at depth below the sea floor are produced by a boring operation.

Sand and Gravel

A large part of the erosion products of the land is carried by rivers into the sea and is there further transported and graded by tides and currents; the coarsest material as beaches, grading off into progressively finer sands, silts and muds in deeper water. Since the world sea level fell by about 100–150 m during the Ice Age there are 'fossil' beaches and river valleys over much of the continental shelf, and coarse gravels can be found scattered at many different places and different depths. As the easily available resources of land areas have been used up, and as alluvial plains have become occupied by expanding urban development or used as valuable agricultural land there has been pressure to develop offshore sand and gravel resources, which are now of major importance.

Both quantity and value of marine dredged sand and gravel is greater than that of any other solid marine resource (it is exceeded in value only by hydrocarbons). In the U.K. mining of offshore resources has exceeded 13 million t.p.a. of sand and gravel since 1968, and according to a United Nations Review the annual U.S.A. extraction rate was 550 million tonnes in 1970, worth $US100 million, and worldwide this resource reached nearly half the total value of marine mined minerals.

Transport costs loom large in offshore exploitation of sand and gravel, and sources near markets are strongly favoured. Water depth is also important economically, since deeper operations involve heavier and more costly equipment and greater lifting costs. Recovery in depths up to 30 m is consequently favoured, so that the activity is likely to remain close to land in the foreseeable future. Both bucket-chain and clam-shell dredging may be used in shallow water (Fig. 3.1).

Fig. 3.1 A bucket dredger (left) cutting a pipeline trench in the Menai Straits, discharging dredgings into a barge alongside. A clam-shell dredging operation is seen in the distance. (Photograph by courtesy of the Westminster Dredging Company, Alton, Hants.)

Extraction of bulk sand and gravel between tide marks using conventional land machines has been carried out in some areas and is a particularly cheap and easy procedure for rapid development of these reserves. It is, however, likely to be open to considerable environmental objection, since on an exposed coast where the material is well washed and graded, its removal from the sedimentary fringe is likely to reduce amenities and to accentuate coastal erosion. It is not a process to be encouraged in countries where preservation of the coastline and its ecological and tourist attractions are ranked as important. Similar environmental considerations apply (although less obviously) to mining within the tidal zone or within the range of breaking waves, as advancing waves stir up the bottom sediments in shallow water, to be returned on the outward flow, and there is a constant interchange between the beach and the shallows which can only be disturbed with adverse results. There is, however, probably little exchange between the coastal deposits and those further offshore, many of which relate to deposition at times of world-wide lowered sea level. Exploitation of offshore sandbanks or sea floor sands is thus not open to the same objections, but any deepening of the sea near land is likely to increase coastal erosion.

Calcium Carbonate

Some countries, for example those with a terrain of acid basement rocks or fine grained alluvium, are poorly supplied with limestone deposits, and historically it has been a common practice in tropical areas to collect modern corals for lime burning and manufacture of mortar and cement. This still takes place in Hawaii and Fiji, but in many places destruction of coral reefs is open to ecological and environmental objections. More recently, exploitation of submarine calcium carbonate deposits has taken place on a large scale. In the Gulf of Mexico, for example, the Mississippi delta is a region very short in exploitable limestones; Iceland is another. In both areas offshore shell beds are mined for cement manufacture and agricultural use.

Extensive areas of the Bahamas Banks consist of fine grained aragonite sand (97% calcium carbonate). Large scale extraction from the Great Bahamas Bank takes place for use in the coastal areas of the U.S.A., with development of the world's largest single mining operation based on an artificial island, Ocean Cay.

Considerations of transport costs and dredging depths, relevant to sand and gravel, apply also to the offshore mining of calcium carbonate, but since the lime-secreting organisms draw their material from sea water this represents a continuously renewed resource.

Production of offshore calcium carbonate reached nearly 20 million tonnes by 1970 in the U.S.A., and is around a million tonnes a year in the Bahamas. In Iceland about 125 000 tonnes of algal debris is mined annually, largely for direct agricultural application.

In an analogous operation, molluscan shells are dredged off the Texas coast for use as a source of magnesia ($MgCO_3$).

Sulphur

Sulphur is a necessary element in the manufacture of fertilizers and in many industrial processes; world consumption is around 30 million t.p.a.

The two main world sources of sulphur are from natural gas, both on and offshore, considered in Chapter 7, and as a secondary mineral in the cap rock of deep seated salt plugs (intrusions of NaCl into the surrounding sediments) particularly in the Gulf of Mexico (Weeks, 1968). Natural gas is a dominant source at present, but the demand for sulphur in industry and for fertilizers will long outlast natural gas resources, and in future there will be renewed emphasis on extraction of the solid mineral.

The extraction system for solid sulphur — the Fraasch process — has been developed on land; it involves injection of superheated water in order to melt out pure sulphur from the mixture of minerals in the highest part of buried salt plugs (Fig. 3.2). This operation has been extended

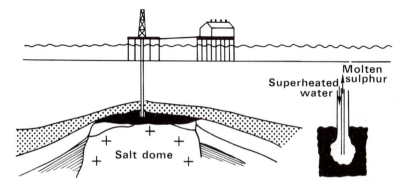

Fig. 3.2 Diagram illustrating the process for the recovery of elemental sulphur from offshore deposits. Sulphur occurs in the cap rock of some salt intrusions and can be recovered by melting out with the injection of superheated water.

offshore since 1960, with the largest of these operations, on the Grand Isle dome, producing 1.5 million long tonnes of sulphur annually. Offshore production is likely to be further developed in line with world demand. Foreseeable activity will probably be limited to exploitation of salt domes in shallow water within 70 km of shore; but it should be noted that the generation of elemental sulphur from natural gas leakage involves complex circumstances of oxidation–reduction, and relatively few salt domes (17 cases of the 350 salt domes on and offshore in the Gulf of Mexico) are known to contain exploitable sulphur.

Phosphates

Phosphorus is an essential element in organic life. Large quantities of it reach the sea from the land as both organic and inorganic detritus, and phosphates are a major requirement of the fertilizer industry, although very little is so far used in the developing countries. One important source is provided by the guano of oceanic islands, a reaction product of the excrement of colonial fish-eating birds with the coral limestone surfaces on which they live. Historically this source has been cheap and of great importance, but the larger deposits have been exhausted and there are now serious environmental objections to the surface destruction which is involved in the process. This is the more serious since the guano deposits largely coincide with the limited agricultural land on oceanic islands.

The alternative source is phosphate deposited in the sea, either on the sea floor as small nodules as it is at present, or preserved from past ages in the bedded sediments of earlier times. The terrestrial deposits of marine origin, uplifted above sea level, have been the main alternative to guano; many of the richest of these have been exhausted but others still have long term reserves. Nevertheless heavy demand has required attention to the modern sea floor, where the mineral phosphorite (calcium phosphate) is locally present in potentially economic quantities as nodules or as phosphatic mud. Phosphatic nodules are widespread, occurring particularly on the western sides of continents, where their distribution tends to coincide with areas of upwelling water, as off Southwest Africa and California (see Fig. 3.3). In these areas they may be abundant in the anaerobic muds of basins, particularly in the oxygen-minimum zone; in other areas phosphatic nodule concentrations are found on shallower shelf areas swept by strong currents, or associated with terrigenous glauconite sands (a common case in fossil examples), as on the Chatham Rise south of New Zealand, and the Blake Plateau off the Florida coast. There appear to have been particular periods in the past when marine phosphate deposition has been especially marked — for example during the Cambrian, Lower Eocene and Miocene — and some of the phosphate suitable for mining on the modern sea floor shelves is a Miocene deposit exposed by deep current erosion.

As with other minerals, marine phosphorite is widely distributed; its economic recovery depends on adequate concentration and, in the long run, on the availability and exhaustion of the more easily won sources on land. Most areas of potentially economic deposits of submarine phosphorite are on the outer margins of the continental shelves in depths of 100–200 m of water. Large-scale dredging techniques for these water depths have not yet been developed, and the profitability from marine phosphate at the moment does not justify the use of the techniques developed for trial mining of manganese nodules in the deep ocean. This is due to the much higher value per tonne of the metal-rich manganese nodules in comparison with phosphate.

Diamonds

Diamonds were formed at depth in kimberlite volcanic rocks, from which they are mined in South Africa. Much richer secondary concentrates of

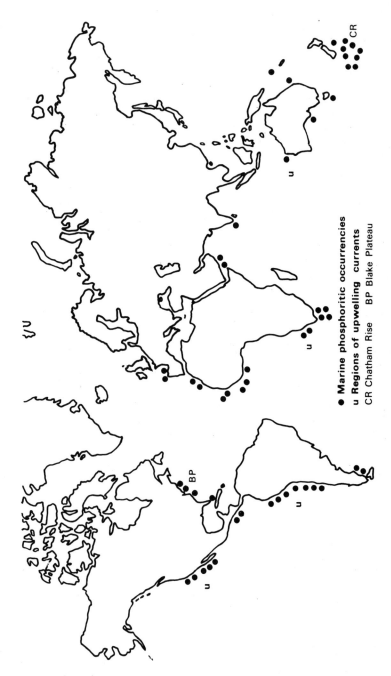

Fig. 3.3 World-wide distribution of sea floor phosphorites. Many of these are associated with regions of upwelling currents.

● Marine phosphoritic occurrencies
u Regions of upwelling currents
CR Chatham Rise   BP Blake Plateau

diamonds are available in alluvial deposits in Namibia (Southwest Africa) and it has been found that these diamondiferous gravels continue off-shore (Fig. 3.4). Dredging was carried out from 1962–1971, but was suspended as costs were higher than on land. There is, however, a long-term reserve of diamond gravels off the west coast of southern Africa available for further development when the richer land sources are worked out.

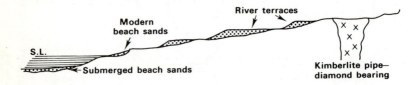

Fig. 3.4 Diagram illustrating the source of offshore diamond bearing gravels.

For exploitation the diamond bearing gravels are stirred up by water jets and brought to the surface by an air-lift hydraulic dredge. However, as with other minerals, there is difficulty in sweeping the surface of the bed-rock where the gems tend to be concentrated. On the mining barge an initial grading stage is followed by a heavy-liquid separation operation and then conventional trapping of the diamonds on a grease belt. Throughput of gravel has averaged about 600 tonnes per day, with production quoted as up to 700 carats of gem-quality diamonds (Mero 1965).

The area of Southwest Africa is one of moderate currents and swell, but is subject to severe storms; in which the intrinsic un-seaworthiness of dredging vessels has been a severe disadvantage. The first dredger was wrecked in 1963 and a supporting vessel lost in 1965. Attention was concentrated on the gravels in bay areas, mostly in depths of less than 20m, but ranging from 10–40m. Costs were some four times higher than those of either the mainland alluvium or the beach deposits so that when the concessions offshore came under the same ownership as the landward deposits, the operations were discontinued.

Barytes

Barytes, in small quantities, occurs widely in the oceans, occurring as concretions in sediments in the Indian Ocean, in Indonesia and off California. Quantities are believed to be below the economic level. Mining of barytes is an operation unique to western Alaska, beginning there with the conventional mining of a small island of barytes rock, and continuing after exhaustion of the material above sea level by drilling and blasting of the sea bed. Production of 122 000 and 93 000 tonnes was recorded in 1970 and 1971 respectively, to a value of around a million dollars a year.

This operation may be a model for other sea bed mining projects in due course, as the search for minerals in the exposed crystalline rocks of continental shelves proceeds.

Dredging Techniques for Bulk Minerals

There are three main types of dredger for lifting solids from the sea floor — bucket-ladder for shallow depths; suction, which may be combined with air-lift at great depths; and wire line dredging using a clam-shell grab (Fig.3.5. See also Fig. 3.6).

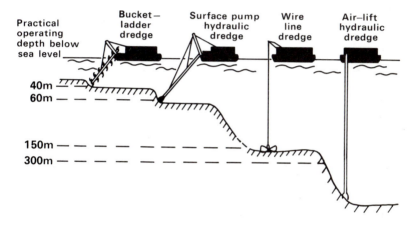

Fig. 3.5 Dredging techniques at different depths. (Modified from Mero, 1965.)

All three systems require that heavy machinery be lowered beneath the support ship, that considerable power be applied to cut or dig into the sea bed, and in most cases that a considerable weight rests on the sea bed. If such a mechanical system is exposed to strong currents or large waves the relative motion of the hull, heaving, twisting and turning and moving relative to the sea floor, results in excessive strains. The deeper the water the heavier the machinery suspended from the ship, and the more likely the ship is to be exposed to the large waves and storms of the open ocean. As a result, there has been no increase in depth capability of open sea dredging for bulk minerals in the last 15 years. The limit has remained at about 30—40 m throughout that time. More and more exploitable deposits have been found within that depth range, but the operational depths have not been extended.

During the same 15 years, as we shall see, the depth capability of oil production has increased dramatically, and some of the techniques developed for the oil industry have been adapted to design suction dredgers

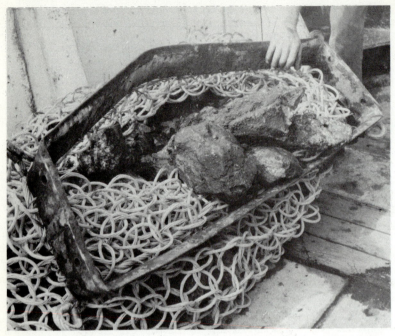

Fig. 3.6 Dredge with rocks from the deep sea floor. (Photograph by courtesy of the National Institute of Oceanography.)

for manganese nodules. The relative failure to develop bulk dredging at outer continental shelf depths underlines the importance of considering the overall economic demand for a resource, and the technical difficulty of extracting it.

4 Metals from the Shallow Seas

Areas exposed above sea level, whether continents or islands, are continuously subject to erosion, millenium after millenium. The products of erosion move to lower levels carried by wind, landslip and water transport; they may halt on the way as river gravels or as lacustrine sediments but sooner or later they all reach the sea, the universal sink, either as detritus or in solution. This is as true for diamonds or the individual mineral grains in a granite as it is for sand and mud, for economic minerals as well as for waste.

Fracture, abrasion and solution reduce the size of the detrital material in the course of erosion and transport, so that the size range of minerals reaching the sea is generally smaller than those on land. The sea has one major advantage, it is constantly in motion; its waves and currents are continuously sorting and grading the detritus which it receives, both as to size and density differences. In consequence of these processes marine sands and gravels may provide rich sources for a wide range of minerals, now being exploited widely round the world.

In addition to the ageless process of erosion and transport from land to ocean a further piece of geological history must be taken into account. Towards the end of the Tertiary Era the formation of continental ice sheets began to abstract water from the oceans, a process which reached its maximum during the Pleistocene and which lowered sea levels, world wide, by several hundreds of metres. Over this period coastal sands and gravels were deposited not along the present coastlines but at varying distances offshore, their distribution depending on sea floor gradients and the extent of local regression. Not only incoming material was accumulated seaward of the present coasts, but pre-existing coastal sands and gravels themselves were subject to erosion and contributed to the accumulation of coarse sediments now permanently submerged to depths of up to 100m. The shallower end of such deposits is now being exploited (down to the economic limit of dredging for particular minerals); they are likely to become increasingly important in the future.

The sorting action of continuously moving marine waters is superb, and provides a high potential for separation of heavy minerals from the much greater bulk of common substances. The sea has, however, obvious disadvantages for dredging operations — most notably the incidence of storms, which dredgers are particularly ill-suited to face and which have resulted in major wrecks and loss of machinery. It also provides a problem which is less acute in land dredging operations, in that the riffling of a sand body during its initial deposition and subsequent disturbance tends to concentrate an important part of the heavy minerals in holes and depressions in the substratum (Fig. 4.1). In land operations such rich pockets may be directly visible or more easily worked by a dredge floating in an artificial lake, but at sea they are difficult to detect and still more difficult to work. With interruptions due to weather and the hazard of

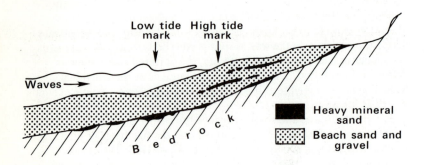

Fig. 4.1 Occurrence of placer deposits containing metallic ores on beaches and offshore sands. The tendency to accumulate in depressions in the substratum provides a problem in dredging. (Modified from Mero, 1965.)

missing rich basal patches, marine dredging is intrinsically less efficient than land based operations, and the difficulties increase the further offshore the operations are located. Nevertheless the continental shelves provide a major potential for future resources, and these will be developed as economic factors permit (Archer, 1973).

Geophysical survey provides a means of locating and defining many submerged ore bearing sands. Magnetometer surveys can directly locate accumulations of magnetic material, the weakness of the signals being offset by proximity to the receiver of the heavy mineral sands. For detailed definition shallow seismic reflection surveys — 'sparker' or 'echogram' type — are used for mapping the geometry of sand bodies, and the location and shape of buried channels (Fig. 4.2). The latter potential is of particularly high importance in following the submerged river channel which are the main source of offshore tin placers in the East Indies.

Gold and Platinum

Gold is unique for its position in international finance; gold and platinum are not only widely used in jewellery but also in the electrical trades, in

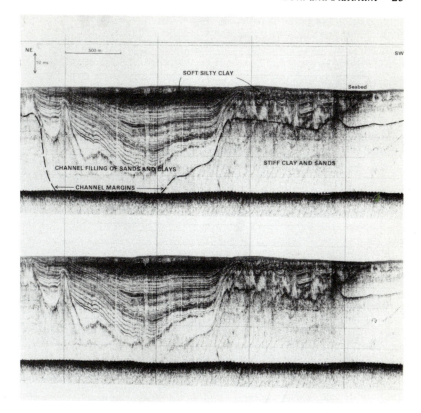

Fig. 4.2 Sparker profile of buried channel in superficial deposits on the North Sea floor. The lower figure shows the basic data; in the upper figure sedimentary contacts are annotated.

chemical industry and in dentistry. The noble metals are extremely resistant to chemical action and they are very heavy, both factors which are conducive to survival and accumulation in detrital sediments, onshore and offshore. They are, for example, reported to occur 'almost without exception' in the beach sands of the western U.S.A. and in fact occur worldwide in sands derived from crystalline basement rocks. Fourteen beaches along the western U.S.A. mainland have been mined for gold and ten more are listed for Alaska. Gold sands also occur in shore sands in Canada, Central America, the Red Sea and other places.

The most famous gold bearing beach sands are those at Nome in northwestern Alaska, which yielded $100 million worth of gold. It is known that these sands continue below sea level, and they have been worked there and in the Norton Sound area some 80 km further east. Other beaches in Alaska have also been worked for gold and, additionally, platinum bearing sands are important in southern Alaska.

In Australia gold bearing gravels occur as raised beaches along the New South Wales coast and have long been worked. Gold bearing lodes in this area are known to continue offshore and there is a potential for subsea mining in the solid rock.

During the late 1960s there was a large investment in prospecting for precious metals on the continental shelves and a number of specialist companies were set up. Many of the activities of this 'boom' era were reported in a journal called *Geo-Marine Technology* which is now defunct. In spite of the enormous enthusiasm generated by the discovery of traces of ore in many localities, very few of the discoveries led to economic projects, and much of the optimism finally drained away. Nevertheless, a technological breakthrough, or a shift in the balance of world supply and demand arising from some dramatic political change, could revive interest.

Thorium

Thorium, a radioactive mineral used in photoelectric cells and gas mantles, has risen in importance with the potential of replacing uranium for atomic power. It is obtained from the heavy mineral monazite $(Co, La)PO_4$ which may contain up to 15% of thorium. It is widespread in occurrence in beach sands particularly in association with the noble metals, for example along the western coast of U.S.A., India and Brazil. The main coastal source is in eastern and western Australia, where monazite is recovered in working the beach sands for titanium minerals. The dredges float in artificial lakes dug into the beach,which fill naturally with sea water. The dredge 'migrates' along the beach sucking up the sand on one side, extracting the monazite and other heavy minerals, and re-depositing the residual sand on the other side. Enormous lengths of beach have been processed in this way, and the industry has been criticized by conservationists. Care is now taken to ensure that dunes and sand ridges are re-planted with protective natural vegetation.

The atomic power-generation potential of thorium makes monazite a strategic mineral, and its occurrence in ilmenite-bearing sands on the Brazilian coast has led to a ban on their commercial working.

Titanium

Titanium is a metal of great importance for steel making and, in pure form, for use in extremely high strength aircraft and rocket components, and cutting tools. A large tonnage of the oxide is used in pigments. It occurs naturally as ilmenite $(FeTiO_3)$ and rutile (TiO_3), hard minerals which are particularly resistant to abrasion and which are worked widely in beach sands. These deposits provide the source of most of the world's titanium. Ilmenite and rutile have been extensively worked on the coasts of eastern and western Australia. Other deposits include California (sands with an average of 75% of ilmenite) and Taranaki in New Zealand where the rich deposits mined in the 1930s contained as much as 40–50% of the heavy mineral. Twenty years later a 10% content was regarded as the limit, but deposits well below 3% content can now be mined economically. Australian production of ilmenite has reached 450 000 t.p.a., but reserves are insufficient to maintain production at these levels without development of submerged reserves, which are likely to be extensive.

Ilmenite has been worked in Florida, in California (Redondo Beach, with 7% of the mineral) and extensive reserves are available in New Zealand (Taranaki beaches on average assaying some 9% of oxide). Rich ilmenite and rutile sands have long been worked in Ceylon, where the balance between exploitation and accumulation is such as to make this a renewable resource, an exceptional situation in mineral development. Ilmenite content of up to 80% and rutile content of 6–10% are reported from this area.

Zirconium

Zirconium is a metal of many uses. It is used in steel making (for purification), in special steels, and as the oxide, zirconia, as a refractory, in abrasives and in paints. Zircon ($ZrSi_4O$) is another heavy mineral occuring in very small quantities in igneous rocks but found concentrated in beach sands, which is the main source of metallic zirconium. The heavy mineral sands of north-eastern Ceylon are a major source with a 6–7% zircon content; zircon is also a by-product of mineral sands exploitation in Australia, India, Brazil and other countries.

Chromium

Chromium metal is used in large quantities for the manufacture of special steels, particularly stainless steel, and in plating. Chromium salts are used in pigments and in many industrial processes. Although chromite ($FeCrO_4$) is mainly produced by conventional mining on land, it occurs as a detrital mineral in shore sands which have been worked in Australia and under conditions of scarcity, due to wartime interruption of supplies, on the Oregon coast of U.S.A.

Iron

Of the many iron minerals which occur in continental rocks, magnetite (Fe_3O_4) and ilmenite ($FeTiO_3$) are sufficiently hard and resistant to solution to accumulate as a detrital mineral. Ilmenite sands are worked for their titanium content, but magnetite sand occurs in much larger quantities available as a major source of iron. The largest operation works magnetite sand in 20–30 m of water in Ariake Bay off the south coast of Japan; reserves are expected to be greater than 16 million tonnes. The ore content is low (with a maximum of about 3% iron) but production is nevertheless at the level of 30 000 t.p.a.

Other large scale reserves are reported off the U.S.A. coasts in Oregon, California and Alaska, and New Zealand, but potential sources are of world-wide distribution.

Tin

For many years the greater part of the world's tin in the form of the oxide, cassiterite (SnO_2), has been mined by dredging operations in artificially flooded pits. Bucket-ladder dredges are floated in pits dug into the unconsolidated sands and gravels which contain the tin ore. Since the water is completely sheltered and still, it is possible to lower the heavy booms of the bucket-ladder dredges into direct contact with the underlying rock, and to scoop up the richest pockets of ore which occur at the bottom of the gravel.

The extension of these operations from landward alluvial deposits to the offshore area was an obvious transition. Submerged tin-bearing river channels cut and filled during the Pleistocene period of lowered sea level are important, and in some cases, for example off Thailand, are worked in 30–50 m of water. Geophysical methods are used to locate and follow the buried channels, which are worked in the Indonesian region by grabs, ladder dredgers or hydraulic dredging. In Thailand the main operations are off the eastern coast of the Malay peninsula; in Indonesia mining is taking place in 20–30 m of water off the tin islands of northern Sumatra. Similar developments are taking place off the Malaysian coast.

Submerged tin gravels have been worked off the Cornish coast, but operations are now suspended. Placer tin deposits are also known off southern Alaska, but world-wide reserves of this kind are believed to be very limited in extent.

5 Metals from the Deep Oceans

A little more than a century ago H.M. Research Ship Challenger discovered that wide areas of the deep ocean floor are littered with manganese nodules – light brown and black, yellowish or reddish, with the same range of size and irregularity as potatoes (1–15 cm). In some areas they are thinly scattered, elsewhere they are described as being as densely packed as a cobbled street. The main bulk of manganese nodules is made up of iron and manganese oxides (Fe_2O_3 and MnO_2), but they derive their special economic interest from smaller quantities of much more valuable metals, notably copper, cobalt, nickel and molybdenum. In total tonnage they represent a vast resource for the future, if practical economic means can be found to mine them from water depths which are often in excess of four kilometres (Archer, 1976).

Many of the nodules show a concentric structure and they have evidently grown *in situ*. Rather surprisingly, they occur only on the sedimentary surface: despite the continuous slow rain of sediment they are not found buried, and it seems that the reducing environment within the sediments converts the oxides to a soluble form which is continuously reprecipitated in contact with the oxidizing bottom water. In line with this concept, many nodules have carious, eroded under-surfaces, but are either smooth or knobbly on top. Beyond that, the origin and development of the nodules remains something of a mystery. Submarine volcanic action could be a source of metal salts, but the distribution of the nodules shows no relation to that of volcanoes or mid-ocean ridges. Terrigenous sources could provide the metals via river systems, but there is no correlation between drainage basins and nodule occurrence. There is, however, an apparent coincidence between richer areas of nodules and those with particularly abundant surface microorganisms (phytoplankton). This suggests that trace quantities of the metals may be secreted by organic life and sedimented either as dead organisms or as faecal pellets, which pass into solution and are subsequently precipitated on the ocean floor in colloid form.

The compositional variation of the nodules, first elaborated in detail by Mero (1965), is a matter of high importance in relation to the economics of mining this resource. Iron and manganese are universally dominant; the usual proportions of iron are in the range 5–15% and manganese

10–30%. More important are the amounts of nickel, copper and cobalt, which together with lesser amounts of lead, zinc and titanium aggregate 1–3% dry weight of the nodules.

In the Pacific, iron-rich nodules tend to occur near the continents, reflecting the early precipitation of this metal in coastal regions; these regions are however generally low in the more valuable metals. High-manganese nodules occur in belts near North and South America; they tend to be high in nickel (an analysis quoted, on dry-weight detrital-free basis, gave 43% Mn, 1.8% Ni and 0.7% copper). More central areas, that is the greater part of the Pacific basin, appear to be high-manganese regions with high nickel and copper; average assays (dry-weight, detrital-free) are given as Mn33%, Ni 1.52%, Cu 1.13%, Co 0.39%, and lead 0.18%. A variation of this situation found on topographic highs in the west central Pacific is a region of nodules high in cobalt, values ranging from 0.7–2.1% and averaging 1.2%, reflecting either unusual source rocks of seamounts or possible abnormally high oxidizing conditions.

These percentages are small, but the total tonnages involved in the whole Pacific are enormous; that of manganese is estimated at 400×10^9 tonnes, cobalt 5.8 billion, nickel 16.4 billion, and copper 8.8 billion tonnes. If only a minute part of these quantities can be economically mined they will extend world resources of critical metals for many centuries.

Other oceans are much less well documented; manganese nodules are known to occur in quantity on the Blake Plateau off the Florida coast, and although their distribution is not well known in the Atlantic and Indian Oceans, which are much younger in origin than the Pacific, these appear to be generally less rich.

For economic exploitation, the regional variability puts a premuim on knowledge not only of the whereabouts of the densest nodule fields, but also of those with the highest values of nickel and copper, which are the primary objectives. This information is now blanketed by strict commercial secrecy. Despite the enormous problems of mining minerals beneath the 4–5 km fluid blanket, five international consortia are now in existence to exploit the manganese nodule resource. These consortia have been working for the last 5–10 years in the Pacific, and have invested many millions of dollars in prospecting and test refining of the nodules.

Location, sampling and relocation of the deposits are the least difficult parts of the operation. Grab samplers which close automatically on hitting the bottom are a simple device, and box corers are used for obtaining undisturbed sections of the sea floor. Sea floor photography is used extensively for assessing nodule abundance; 30 000 photographs are reported to be available to one organization alone. Television cameras are used alone or in conjunction with dredge equipment to observe the sea floor.

Even in the deepest ocean the floor is not flat; slopes of 6° occur and the surface is broken by occasional outcrops of ash or basalt which present a potential hazard to large scale dredging equipment. Fully detailed bathymetric surveys are thus a necessary preliminary to choosing a mining site and for accurate location of the dredge-head. Nodule distribution may be very localized and patchy, and an acoustic beacon network on the sea floor is necessary for accurate location. Satellite navigation is increasing in accuracy to a point where a vessel can be located to within a hundred yards, but equipment lowered overboard is unlikely to sink straight down, being affected by different currents at different depths, and surface and bottom positions are hence largely separate problems.

Of the five consortia actively experimenting with nodule collection, one is using a continuous bucket-chain device and the others various hydraulic systems. The bucket-chain is laid in a wide loop on the sea floor, trailing behind the mother ship, and sweeps a belt perhaps 100m wide as the ship steams slowly forward, lifting up to 3000 tonnes per day. In practice, it has been found extremely difficult to prevent the long bucket chain from tangling under differential current drive, and experiments using a pair of ships are said to be more successful. The objection is also advanced that the buckets may quickly fill with the soft sea bed clay and fail to collect many of the nodules. The principle of deep ocean mining for metalliferous nodules is shown in Fig. 5.1.

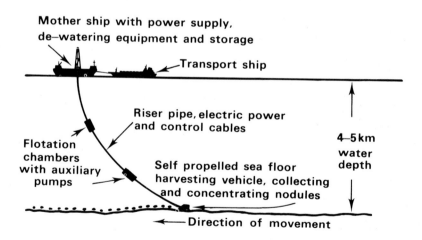

Fig. 5.1 Diagram illustrating the principle of deep ocean mining for metalliferous nodules.

The sea floor hydraulic collector systems are either self-propelled, using electric power, or towed behind the mother ship. The latter are currently favoured, since traction is uncertain on a soft bottom and inconveniently high power is required in self-propelled models. The

method of lowering the dredge head to the sea floor has been developed
from the techniques of the oil industry and the Deep Sea Drilling Project,
where several kilometres of drilling pipe are assembled length by length,
screwed together, and lowered progressively from a derrick. The derrick
itself is mounted on a colossal stabilized platform at the centre of the
drilling ship suspended on gimbals, and compensated to counteract the
wave action. In manganese nodule mining the pipe is used not for drilling,
but for pumping up a mixture of water and nodules. It turns out that
the natural flexibility of the great length of pipe, combined with special
articulated joints near the bottom, allows the forward motion of the ship
to be transmitted to the dredge head on the bottom without breaking
the pipe.

Nodules may be collected directly by a 'vacuum cleaner' device or
heaped in windrows by waterjets and scrapers, before being brought to
surface in a separate operation, possibly by a batch lift system. Some
process of concentration of nodules, as against sediment, is desirable
on the bottom to give a nodule concentration of perhaps 15% in the
slurry (Smale-Adams and Jackson, 1978), but there is danger of loss
of finer grained or friable nodular material in this process.

The collector head has to be instrumented for sensing devices, for
steering around obstacles and for operation of pumps. The sensing devices
include the monitoring of the rate of mass flow, vehicle speeds, heave
and pitch, depth and tow force. In total the head operation requires
several hundred kilowatts of electrical power. Cables, motors and switches
have to be capable of withstanding the enormous pressures of 5 km of
water without failure.

Hydraulic lift systems are under development; these are likely to
depend on lightening the top of the column in the lift pipe by air injection
(Use of hydrocarbons is an alternative, but is open to environmental
objections.) The support of the delivery pipe, particularly if large enough
to provide for air-lift, itself presents considerable design problems, and
an output of 10–15 000 tonnes per day requires 4 MW of power simply
to raise the nodule load against gravity. Much of the required technology
for lifting and concentration is new and is still under development. It is
ironic that the huge lifting vessel Glomar Explorer was developed under
the cover story that it was going to be used for manganese nodule mining,
and was in fact used by the American Navy to lift a crashed Russian
submarine from the depths of the Pacific. After some years of delay
and speculation, this unique ship has now been converted for genuine
manganese nodule recovery.

The mining vessel will have to be highly specialized, and it is expected
that it will stay on station for 3–4 years at a time, shipping the recovered
mineral back by ocean transports. Apart from including a major elect-
ricity generating plant it must have sophisticated separation equipment
and the capability of supporting the lift pipe with its heavy load, support

which must not be subject to the vertical ship movement. The towing forces are calculated to be an order of magnitude higher than those of a normal ocean-going tug. The technical problems are enormous, but so also is the value of the prize if success can be achieved.

It may well be impractical to carry separation of the nodule concentrate to a point where it is effectively dry, and transport of wet or thixotropic material will then require compartmented transport ships to cope with the danger of instability in the event of bad weather. A further environmental link arises in the potential disturbances to benthic organisms due to the clouds of fine sediment generated during mining and washing at sea, affecting (for example) light penetration, but these effects are not expected to be on a large scale. At least one of the consortia has conducted extensive tests, dumping samples of bottom mud from dredging operations overboard and towing instruments through the dispersing plume to measure the concentration of sediment, the rate of sinking, the rate of spreading, and the amount of light absorbed. They concluded that the plume would not be an environmental hazard.

An appropriate size of concession for a viable mining operation, sufficient to provide 20 years or more of working and to be economically profitable, is considered to be 50 000 sq km (20 000 sq miles). Despite the vast size of the oceans there are relatively few areas as large with a nickel and copper content of 2.25% or better; estimates based on samples from two thousands sites vary from as few as 28 to as many as 100 areas with this degree of richness. In consequence there is likely to be strong competition for suitable concessions, and there is a high degree of commercial secrecy about the locations where the present international consortia are exploring.

Despite the enormous overall reserves of manganese nodules, the difficulties of recovering them will limit exploitation to the very few advanced nations who are able to produce the resources of technology and capital necessary to achieve economic success. The variation in abundance and grade will require a high degree of precision in exploration; the environmental problems of weather, water movement and navigation on the surface unrelated to a dynamic situation 4–5 km beneath, the techniques necessary for collecting friable nodules from a fine grained incompetent substratum and conveying them first to surface and then to port add up to a formidable task. Added to these factors is the political element, so pervasive as to be effectively part of the environment, with international agreement on the law of the sea (and hence on exploitation rights) still lacking, with the 115 underdeveloped nations majority ('the Group of 77') resisting development by the industrial nations. Interests in the U.S.A., Britain, Germany, France, Belgium and Japan have formed consortia to develop mining methods (Harris, 1978) and to provide the billions of dollars of capital required to make available these resources of the deep ocean.

The first full scale operations have nevertheless already taken place (in 1978) and it is to be expected that copper and nickel from the Pacific will find their way into world markets soon after political problems are resolved.

Politics and economics however remain the determining factors. Nickel, copper and cobalt are strategic metals, and the advanced countries depend almost entirely upon imports. Provided that world order, political alliances, and price structure permit imports from countries with land deposits, there will always be prospecting and discovery of new low grade ores on land in competition with the nodules at sea. It is this balance, and changes in it, which will determine whether or when manganese nodules become a major source of metals for the western world.

6 Minerals from Solution

All erosional detritus ultimately reaches the sea; most of it settles on the bottom as sediments, some of it remains only briefly in solution and is then precipitated as solid minerals (e.g. iron salts and phosphates, mentioned above). A major part remains in permanent solution. Among the dissolved materials sodium chloride is dominant, with lesser quantities of magnesium sulphate and calcium salts, but at least 60 other elements are present in measurable quantity (Mero, 1965) and from the occurrence of rare elements in organic matter it may be deduced that every known element is probably a sea water constituent. Sodium chloride, magnesium salts and bromine are commonly recovered; extraction of iodine, potassium, calcium sulphate and ultimately perhaps gold and silver are practicable.

Water

Before considering the availability and recovery of minerals by chemical processes, it is as well to remember that the universal solvent — *the water* — is also an economic mineral and is itself recovered from the sea in large quantities on the coasts of water-deficient lands — in Kuwait, Abu Dhabi, Saudi Arabia, Malta, the Canary Islands, the Bahamas and even in less arid countries such as the U.S.A., Mexico, southern U.S.S.R. and the Netherlands. An annual output of 284 million tonnes was quoted by McIlhenny in 1975, and the amount increases every year.

Recovery of fresh water, by distillation, from the sea has a long history, documented at least as far back as Roman times. The greater part of modern production relies basically on the same process in more sophisticated form, using multi-stage flash distillation. In principle, the operation is simple; in practice, there are problems, including scale formation on heating surfaces from the dissolved solids, corrosion due to chemical action (ameliorated by the prior removal of dissolved oxygen) and the liberation of carbon dioxide which impedes thermal transfer and also has a corrosive effect.

Other processes include solar evaporators, used for more local supplies, freezing processes (ice formed is largely salt-free) and the use of electro-dialysis with semi-permeable membranes. These latter processes avoid some of the problems of distillation but are less economic for large scale output and are less widely used.

Fortunately, some of the areas most needing water from the sea also have abundant energy in the form of petroleum, and it is in these cases that sea water distillation provides a major part of the requirements of the population.

Salt

Extraction of common salt, halite (NaCl) from sea water has a history extending back into pre-historic times, documented by the Chinese from 2200 B.C. and operated, for example in Britain (Lincolnshire), in Iron Age times. Salt is widely produced on coasts by solar evaporation, most effectively in desert areas with low humidity, low rainfall and high evaporation rates but also in temperate countries.

The operation involves a series of evaporating ponds (Fig. 6.1), the initial basins filled either by high tide or by pumping (windmills may be used for power) and serving for settlement of sediment, iron compounds and calcium carbonate and for initial concentration. In the second stage

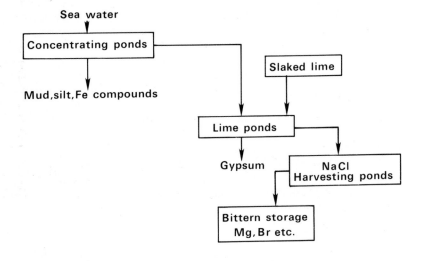

Fig. 6.1 Diagram illustrating the principle of salt (NaCl) recovery from sea water. (See text for details.)

(lime ponds) calcium sulphate is precipitated. The third stage consists of final harvesting ponds in which evaporation is taken to a point when most of the sodium chloride crystallizes out, the liquor being pumped out when a density of about 1.26 is reached to prevent deposition of magnesium and potassium salts. Where chemical facilities are available the

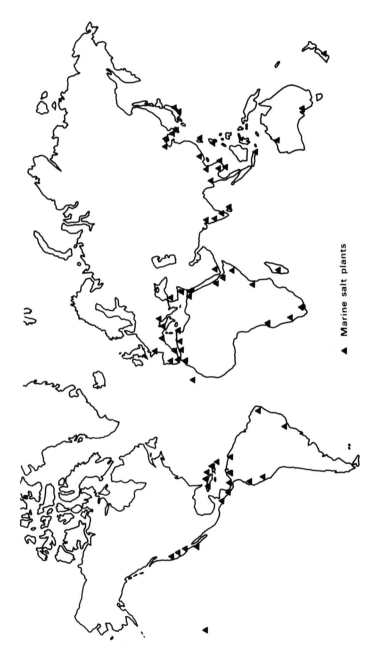

▲ Marine salt plants

Fig. 6.2 World-wide distribution of salt recovery plants.

final liquor, known as bittern, is used for production of magnesium salts and bromine. The efficiency of the evaporating process can be increased in sophisticated operations by adding a green dye to the sea water; this reduces reflection of the sun's rays and adds considerably to energy absorption. The distribution of salt recovery plants is shown in Fig. 6.2.

Potassium

Potassium is an essential ingredient of fertilizer, and sea water contains vast tonnages of potassium salts. An early source was the seaweed industry, since marine algae concentrate potassium and other elements which can be recovered from the ash. The ash of the brown seaweed *Laminaria* contains up to 33% of potassium carbonate. Potash is also available from the bitterns, the concentrated brine residuum of sodium chloride manufacture, but there is no significant commercial recovery since recrystallization of these highly soluble salts, based on evaporation to dryness, is not economic in competition with mined sources. Alternative methods of recovery involving insoluble potassium amines and other compounds have also been devised (McIlhenny, 1975), but the potassium of the oceans remains as a reserve for the future.

Bromine

Bromine was initially identified in the bitterns of salt extraction pans in southern France, and the major world source is still sea water. Bromine was originally discovered in 1811 in the ash of seaweed, and this served as the main source for a century, an industry based mainly in northern England.

Demand expanded enormously with the development of the internal combustion engine and the requirement for anti-knock petrol together with tetra-ethyl lead. A major part of bromine production involves direct precipitation from seawater by treatment with aniline and chlorine. The bromine is separated as insoluble tri-bromoaniline and released by treatment with a mixture of sulphur dioxide and air. This operation depends of course on a consistent supply of untreated water, and the initial plant was mounted in a ship. Subsequently, coastal locations in areas with strong steady currents have been used. There are several plants on the U.S.A. coasts, providing some 80% of national requirements, and in Britain bromine is extracted by this process at a coastal plant at Amlwch in Anglesey. Figure 6.3 shows the distribution of recovery plants.

Iodine

Iodine was originally discovered in the ash of seaweed, of which *Laminaria* has the facility of concentrating the element from a level of about 0.05 ppm

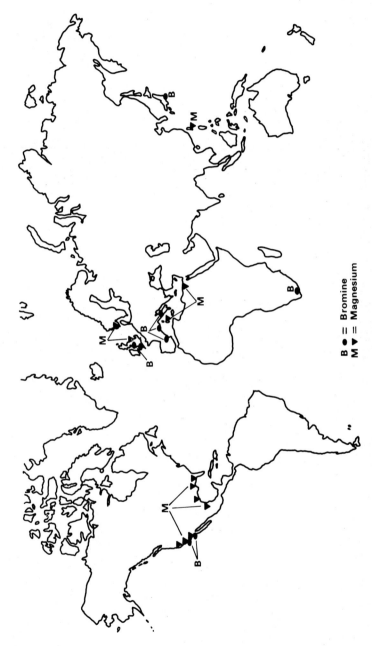

B ● = Bromine
M ▼ = Magnesium

Fig. 6.3 World-wide distribution of plants recovering bromine and magnesium from sea water.

in sea water to 0.5% on a dry weight basis, i.e. by a factor of about a hundred thousand. During the last century this source was vigorously exploited in Scotland, but the industry died away in face of much cheaper production from the Chilean nitrate deposits. The case is interesting, however, as an illustration of the potential of organisms for collecting materials present in very low concentrations in sea water, a potential which might be exploited in the future (as noted below).

Magnesium

Magnesium is required in quantity as a light metal and for a wide variety of chemical and pharmaceutical products. Compounds of magnesium are an important constituent of both continental rocks and sea water; and although concentration in the oceans does not compare with that of easily mineable dolomite, nevertheless marine sources are important for production of high purity magnesium compounds and the metal itself, the process involving reaction of sea water with either lime or calcined dolomite from land sources (Fig. 6.4).

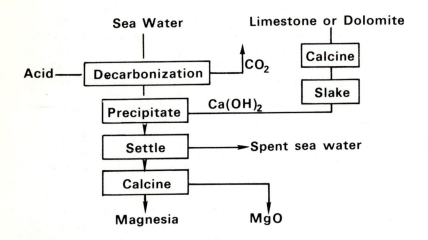

Fig. 6.4 Flow diagram showing magnesium recovery from sea water. (See text for details.)

Sea water contains about 0.13% of magnesium, and (as with bromine extraction) a site favouring continuous change of water at the intake end is clearly essential. The sea water is mixed with a slurry of slaked lime or calcined dolomite, the reaction producing a precipitate of magnesium hydroxide; calcium salts are separated out and high purity magnesium salts or the metal recovered by electrolytic methods.

The process was first used in Britain (Armstrong and Miall, 1946); with variants it is now the main source of magnesium in the U.S.A. (See Fig. 6.3 for distribution of recovery plants.)

Gold and Other Rare Metals

The concept that the ocean waters contain large amounts of gold has led to many attempts at recovery. Original estimates were far too high, but the six million tonnes now thought to be present represent an extremely small concentration — four millionths of a milligram per litre. Other valuable metals such as mercury, uranium and the rare earth elements are present in much larger concentrations, but none of them approaches a level where extraction by chemical means is economically practicable at the present time. The U.K. Atomic Energy Authority conducted experiments in the early 1960s on the absorption of uranium on various titanium compounds, and, whilst surprisingly efficient, they did not justify economic exploitation.

Certain organisms however have a remarkable propensity for accumulating metals present in trace quantities — fish accumulate lead in their bones (average concentration in sea water three hundredths of a milligram per litre) and tunicates are notable for collecting vanadium (two thousandths of a milligram per litre). Concentration of pollutant mercury in Tokyo Bay by fish has led to the dreaded lethal Minamata disease. If the mechanisms of such selective absorption can be understood there might be a future possibility of applying them to other and more valuable metals.

As with other aspects of mineral recovery from the seas and oceans, a most stimulating compilation is provided by John L. Mero in *The Mineral Resources of the Sea* (1965). The book was written in a period of economic and technological optimism which seems strangely dated nowadays. Then it seemed impossible that the spin-off from space technology and the race for the moon must surely provide us with the simpler technology to exploit the entire floor of the ocean and the sea water itself. But all sorts of things have gone wrong in the last decade. Experience has shown that many of Mero's costings and predictions were over-optimistic, even allowing for inflation. Minerals have proved to be in lower concentrations, more widely scattered, and harder to recover than expected, whilst new large-scale operations on land have enabled lower and lower grade ores to be exploited profitably. Paradoxically, while the rapid increase in the price of oil since 1973 has enormously boosted the offshore oil industry, it has produced a general slackening of the world economy so that other expensive and speculative ventures have been abandoned. In this more cautious atmosphere many marine extraction projects have been scrapped. However, in spite of these provisos, Mero's book remains a unique inventory of potential marine resources. A modern account of extraction of the main minerals is provided by W. F. McIlhenny in *Extraction of Economic Inorganic Materials from Sea Water* (1975): this emphasizes the physico-chemical problems and the industrial processes in use. Both sources have been drawn on extensively in preparing this summary.

7 Oil and Gas from the Shallow Seas

Hydrocarbons form a major part of the world's energy supplies. They have been the subject of intensive search for a century, to the extent that most of the land areas have been thoroughly explored and their potential very largely exploited. (The chief exceptions to this generalization are in Arctic regions, where remoteness and extreme problems of climate and subsoil have delayed development.) The natural continuation has been to move offshore into sedimentary basins on continental shelves, a move which has largely taken place since 1950. The continental shelves are shown in Fig. 7.1. This has now resulted in at least partial exploration of nearly all the world's offshore basins, with large scale discoveries for example in the Gulf of Mexico, off Western Australia, off West Africa and in the North Sea.

These operations have involved moving into a new and hostile environment, with problems of weather, sea conditions and sea floor stability; problems which have led to major developments in exploration and engineering technology. They have led also to deeply felt concern about this new source of marine pollution, added to the existing problems of sewage and industrial discharge into the seas. Some of the factors concerned are outlined in the following paragraphs.

Palaeoenvironment and Distribution

Oil and gas are mainly generated and accumulated in marine sediments. The physico-chemical circumstances both of their generation and of their survival are narrowly prescribed. If the organic-rich sediments are insufficiently deeply buried they do not reach temperatures sufficient for oil generation; if they are buried too deeply, only gas is produced. Survival of economic quantities depends on accumulation in a suitable reservoir rock, trapped by an impermeable seal to prevent (or hinder) loss by seepage; on a history of subsequent disturbance which does not break the seal and permit loss of the accumulation, and on subsequent maintenance at depths where the temperature is not sufficiently high to crack oil to gas or, at a further stage, so severe as to crack gas to carbon.

This complex and critical history, from the initial accumulation of an organic-rich mud to the eventual availability of a gas or oilfield for

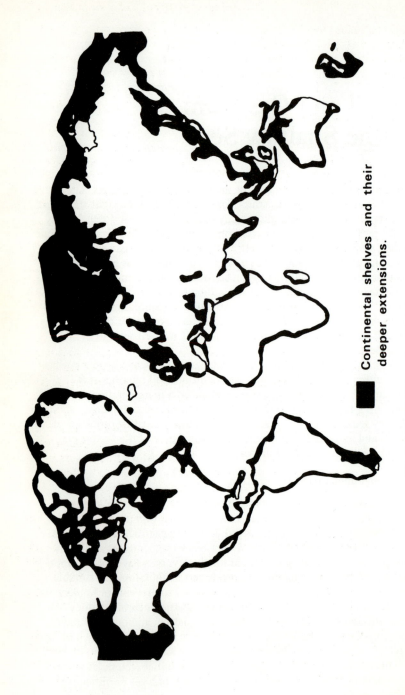

■ Continental shelves and their deeper extensions.

Fig. 7.1 World-wide continental shelves.

exploitation, is most commonly found in the marginal areas of continents where newer (Mesozoic and Tertiary) sediments have accumulated to thicknesses of a few kilometres and have remained relatively undisturbed since their formation. Areas of ancient rocks have mostly been too much fractured, heated and metamorphosed for the survival of fluid hydrocarbons, and areas with only thin sediments do not meet the requirements of suitable past temperature regimes.

In the earlier stages of offshore exploration some geologists held that only those continental areas adjoining known petroliferous provinces on land could be considered prospective, and major areas of the world were consequently largely neglected. Australia demonstrated how unwise this concept was, with its exposed metamorphic rocks now known to be surrounded by submerged sedimentary basins. It also provided a reminder that the shape of some continents is structurally controlled; that there may be a sharp change along a coastal region, with the modern seas or oceans coincident with areas long subject to subsidence, and hence to heavy sedimentation. On a smaller scale the northwest European continental shelf is a parallel case, with the continental margin and British Isles largely reflecting the distribution of ancient blocks which are separated by partly flooded subsiding basins, of which the North Sea is the most important so far (Fig. 7.2). Whether the other basins have comparable potential for hydrocarbon production is as yet unknown.

It has to be emphasized that continental shelves are far from uniformly endowed with hydrocarbon riches. The submerged shelf west of the Hebrides represents an extreme case on the negative side, since the sea is floored by metamorphic rocks. But sedimentary basins themselves may be hydrocarbon rich or nearly barren — as exemplified by the basins around Africa, which nearly all contain important oilfields on the west coast but are virtually barren on the east coast. This case may reflect the long-term effects of oceanic circulation on the organic content of the sediments, upwelling of deep water and such phenomena as 'algal bloom'; it may be that a comparable contrast in hydrocarbon richness or poverty will be found to be general between the coasts of other continents also.

There is another general asymmetry of the continents which may affect petroleum occurrence. Active continental margins near to subduction zones tend to be mountainous and are often areas of recent uplift. Passive margins are in general of lower relief with older eroded mountains, and the offshore areas and continental slopes are generally subsiding slowly. As a result the largest continental rivers tend to drain across the width of continents from the high mountains near the active margins and deposit their sediments on the passive margins.

This kind of situation makes nonsense of statistical attempts to estimate the world's ultimate petroleum reserves from an assessment of the volume of sediments in different areas. As an example may be quoted the crediting of the Canadian Arctic with 40 billion barrels of recoverable oil reserves

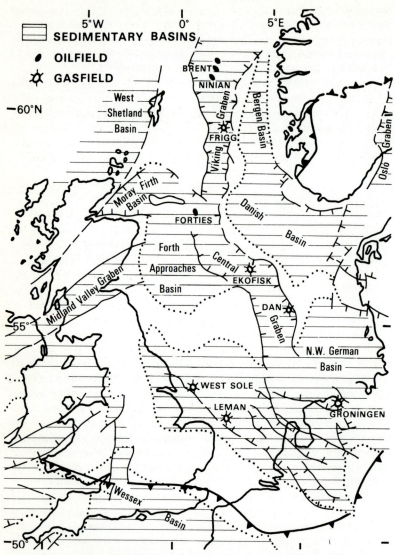

Fig. 7.2 Blocks and basins in the north-west European continental shelf. Major oil and gas deposits are limited to the basinal areas with thick sediments.

before drilling; the figure achieved after dozens of deep exploration wells is little more than one percent of this estimate. On present evidence a final score of two billion barrels for this region, land and sea, would be highly optimistic.

The continental margins cannot *all* be considered to have a high potential for hydrocarbon discovery, but the circumstance that they contain in total a vast volume of sediments which include suitable source rocks, reservoir and cap rocks, and that most of these sediments have passed through the temperature regime appropriate for generation of hydrocarbons, makes them attractive areas for exploration. One important variant between different areas is the distribution, quality and thickness of reservoirs, commonly porous sandstone, less frequently limestone, which bears not only on the size of hydrocarbon accumulation but also on the rate at which the fluid can be extracted. High production rates may offset the greatly increased cost of establishing production beneath deep water; low productivity may inhibit activity except in very shallow conditions.

Rivers and subsurface currents however drop the coarser part of their load (boulders, pebbles and sand) before the finer muds, and, with important exceptions, there is consequently a tendency for fine grained deposits to predominate further offshore, with progressive deterioration in reservoir quality in deeper waters. The chief exceptions are found off large rivers where coarse material may be carried far seawards in submarine channels and canyons, and from the activity of turbidity currents which may carry coarse sediments along deep depressions parallel to coasts. Channel sands are however difficult to find, and the sands deposited by turbidity currents are often poorly sorted and of indifferent quality as reservoirs.

It follows that in any single formation the better sands are found near the sources of sediment — particularly on the proximal parts of continental shelves — but that there is a progressive deterioration in reservoir quality towards the oceans. This is a feature traceable in marine sediments on land and is important in finding the most favourable productive areas; it is still more important offshore where the expense of exploration is much greater. However, the vagaries of transgressions and regressions of the seas during geological time have resulted in sands being deposited much further out — during marine regressions — and blanketed by finer grained sediments (clays) when the shoreline lay further onto the continent. Thus off western Australia good reservoir sands were deposited on the edge of the present continental shelf in the early Mesozoic (Triassic), and these were covered by late Mesozoic (Jurassic and Cretaceous) deeper water shales and clays which form a seal on gas bearing sands.

The range of reservoir type is considerable. The better gas sands of the Southern North Sea were deposited as wind-blown coastal dunes in Permian times, and are now deeply buried. Reservoirs of lesser value are provided by thin finer grained water-laid sands further from shore. The main Jurassic reservoir sands in the Northern North Sea were deposited under deltaic conditions, as were the sands of offshore Louisiana. The later North Sea sands (Palaeocene) which contain the Forties oilfield are of turbidity type, as are many of the major sand reservoirs in California, but turbidites elsewhere around the Pacific (e.g. Southern Alaska, New

Zealand, Japan) have proved disappointing or non-productive. The rich oilfields of the Middle East are almost all in limestones, onshore as well as offshore, with the major exception of Kuwait where the oilbearing Burgan sand is of Coal Measure type. The excessive richness of the Middle East — with 51% of the world oil reserves in an area 500 x 800 km — reflects not only large structures, excellent cap rocks and very thick reservoirs but also specialized conditions of source rock development, perhaps deposition in a partly enclosed oceanic gulf, which favoured an abnormal original richness in organic content of both shales and limestones. But it is an area which is now largely drilled up, both onshore and offshore, and we see no likelihood of finding a region of comparable richness and high production elsewhere.

These comments apply to the continental shelves. The situation on continental slopes and continental rises of the deep ocean is a different matter which is discussed in the next chapter.

The Process of Exploration

Environmental factors are a dominant aspect of offshore activities: most obviously in dealing with the problems of weather, sea conditions and precise location far from land; less obviously in relation to sea floor circumstances which affect seismic survey, foundation and pipeline construction, and the drilling problems which arise in deep penetration of the sea floor rocks.

In line with the development of the drilling capability our understanding of the structure of continental shelves has progressed from an assumption that these are a simple continuation of the land geology — which applies to the nearer parts — to a realization that the edges of continents have a long history of development and are of two distinct types.

The margins of the Pacific have developed in association with subduction (under-riding) of the ocean floor, either under the continental coastal mountain ranges, such as the Andes, or beneath the marginal seas such as the South China Sea, Yellow Sea, Japan Sea, and Sea of Okhotsk. Subduction areas have so far not proved to be environments suitable for major oil accumulations, but the marginal seas trap considerable quantities of sediment, and these contain economic oil deposits off Indonesia and China. In contrast the Atlantic and Indian Oceans are of tensional origin (opening by breakups of an original continuous area of continental crust) and heavy deposition took place in rifts and subsiding basins, an arrangement which has fostered the environmental conditions necessary for the generation and accumulation of hydrocarbons. Extrapolation from the known land geology consequently has sharp limitations, for these special factors apply particularly to the marine regime.

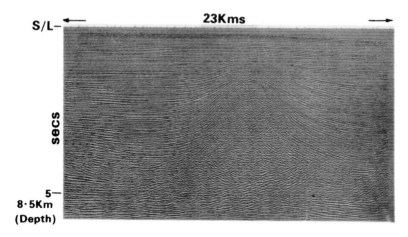

S/L–

23Kms

secs

5–
8·5Km
(Depth)

Fig. 7.3 Seismic record showing deep structure off the South American coast.
The shallower rocks have been bulged by flowage of deep seated salt beds.
(Photograph by courtesy of The British Petroleum Co. Ltd.)

The sea surface provides no clues to the underlying geology, and the
sea floor itself is commonly blanketed by recent sediments. Geophysical
methods are essential from the outset for definition of the deep structure.
The general distribution of sedimentary basins and intervening areas of
shallow crystalline rocks (as in the North Sea, shown in Fig. 7.2) can
initially be plotted by using gravity and magnetic surveys. Low gravity
values indicate major thicknesses of sediments, or salt, while a complex
pattern of magnetic anomalies indicates shallow basement rocks with
their multiple intrusions. Gravity and magnetic surveys however provide
only a broad initial picture, since they sum up the totality of many different effects, and in consequence provide only an approximation to detailed
structure.

Detailed knowledge of deep structure comes from seismic surveys
(Fig. 7.3) — a process of echo sounding of surfaces of the bedded rocks
which provides information on folding and faulting of the deep strata, on
the occurrence of buried limestone reefs, on lensing sands and on the
general lithological sequence, In principle the process is simple, but the
instruments pick up data relating to many scores of separate rock surfaces
beneath each recording point; the critical time measurements may be
distorted by local velocity variations in the shallower rocks, and the
whole recorded sequence of sonic data is overlaid by reverberations in the
sea water column and in the shallower rocks. The important deeper
surfaces provide relatively weak signals, and the ability to work out

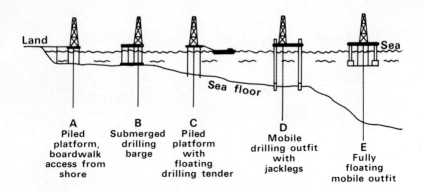

Fig. 7.4 Progress of drilling techniques into deep water.

structure at progressively greater depths has depended on increasing sophistication of the analytical methods. It is now normal for data to be collected in digital form to provide for subsequent computer analysis. Definition of structure down to three thousand metres is now conventional; under favourable circumstances penetration may be more than twice this depth.

Geophysical surveys however cannot provide answers to the final question — whether hydrocarbons are present in economic quantities. For this there is no alternative to drilling holes through the strata, surveying (by physical probes) the nature and fluid content of the rocks penetrated, recovering fluid at surface and measuring potential hydrocarbon production by flow tests.

Offshore Drilling and Production

In some areas the progress from land to offshore exploration was a simple matter of stepping out from existing oilfields onshore or following structural trends into shallow water, as in Louisiana and in the Caspian at Baku. Early exploration was from piers; the next stage was to use grounded drilling barges; jack-up platforms with extensible legs resting on the sea floor followed for depths of 60–90 m (Fig. 7.5). Beyond this semi-submersible floating drilling rigs or drill ships are used (see Fig. 7.4).

The ability to drill exploration wells in deeper water has consistently kept ahead of the possibility of developing oilfields with their complex wellheads and collecting systems, and industry now has outfits with fully developed safety systems capable of drilling in 2000 m of water. Production, however, is still largely limited, partly by economics, to water

Fig. 7.5 'Adma Enterprise' drilling in shallow water in the Persian Gulf. (Photograph by courtesy of The British Petroleum Co. Ltd.)

depths of 200–300 m (Fig. 7.6). Even if prospecting and wildcat drilling reveals large quantities of oil in deep water, the cost of production over many years may be prohibitive. There is the cost of a large permanent platform from which many wells will be drilled, or to which submarine completions will be connected, and from which maintenance of the oilfield will be carried out. Most expensive of all may be the laying of a

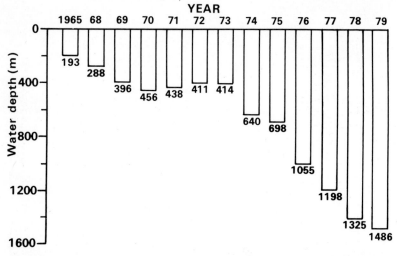

Fig. 7.6 Maximum water depths in offshore drilling operations.

pipeline 100 km or more long in water depths of several hundred metres (see Figs. 7.9, p.55 and 7.12, p.60).

The drilling of an exploration hole requires a drilling outfit in position for only weeks or months, but production facilities require provision of a platform which will stand safely in all weathers and sea conditions for twenty or thirty years. The earlier procedure was by the use of piled platforms on a basis of one per well, connected to a central gas separation platform which fed the pipelines to land (Figs. 7.5 and 7.7). The second stage involved multi-well platforms, from which up to 30 wells may be drilled on a radiating pattern to drain a reservoir over several square kilometres (Fig. 7.11, p.58). This is now the conventional procedure in most offshore fields, limiting costs and minimizing obstruction to fishing and navigation.

The majority of multi-well platforms are of steel construction; the main structure is built on land and floated out using temporary buoyancy tanks (see Fig. 7.8).

Alternatively, very large concrete structures are being used; similarly constructed at the coast and floated out to the oilfield site. (Encouragement to use concrete was available from the experience of offshore forts constructed in the First World War, which have survived intact with no serious deterioration, despite the effect of the marine environment, for half a century.) Several designs have been built and successfully placed in position; the largest of these — and the largest man made movable object so far — is the central platform of the Ninian oilfield in the Northern North Sea, weighing six hundred thousand tonnes.

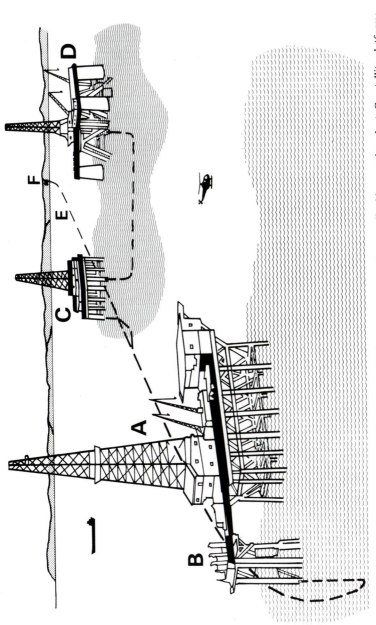

Fig. 7.7 Diagram showing the layout of a North Sea gas field. A, central gas well; B, gas-liquid separating plant; C, satellite platform; D, drilling rig sinking additional wells; E, gas line to shore; F, coastal landing point and further separating unit.

Fig. 7.8 Jacket section of an oil production platform (Graythorp I) being lowered to the sea bed in the Forties oil field in the north of the North Sea in 1974. (Photograph by courtesy of The British Petroleum Co. Ltd.)

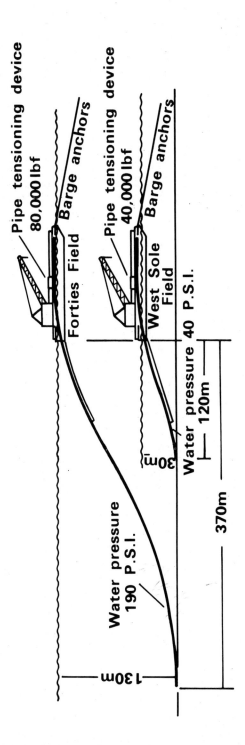

Fig. 7.9 Diagram illustrating the change of scale of pipelaying operations with increasing water depth.

The production platforms are required to carry rigs for drilling the producing wells and for subsequent work-over operations; they carry valve systems for control of hydrocarbon flow and for emergency action; they carry complicated plants for separating light hydrocarbon fractions from the crude oil and pumps for driving the oil through the buried seabed pipelines to shore, perhaps many miles distant.

The machinery, accomodation and power plants are assembled on land into large modules weighing hundreds or even thousands of tonnes, floated out to the platform on barges, and then lifted in position by special cranes (Fig. 7.10). The original emplacement of the platform and the loading of the modules both require at least moderately good weather and sea conditions, and the prediction of calm spells for these operations is an important aspect of environmental forecasting.

Environmental Problems and Pollution

The engineers concerned with organizing and effecting the extraction of offshore oil are well aware of the effects of the environment, of the forces of wind and waves, water temperatures which may be low enough (in the North Sea) to cause embrittlement of some steels, and the ever present corrosion of almost all materials continuously subjected to salt water and sea spray. Advance knowledge of the physical conditions is of high importance in design of structures — it has been said that the difference in estimate between 27 and 29 m for the maximum wave height ('the hundred year storm') in the Northern North Sea involves around a million pounds more or less to the cost of an individual platform. Much of the relevant data on wind, waves, currents and tides there had been collected by such organizations as the Institute of Oceanographical Sciences (particularly for British seas), and the Meteorological Office, and these were critical in engineering planning. In the nature of things however data are largely collected along coasts and this has had to be supplemented by specially designed recording buoys to provide direct information on conditions far offshore (Fig. 2.5, p.8).

Offshore oil platforms, both exploratory and production, have been in use for several decades now, and the platforms themselves provide the basis for detailed observations and measurements of environmental conditions, as well as measurements of the response of the platforms to the environment. Meteorological recordings are now made as a matter of routine from platforms, and some are instrumented for oceanographic observations, measurements of currents, waves, and water temperature. If the platform is also instrumented to measure and record its oscillations and vibrations and bending, then the degrees of movement and possible

Fig. 7.10 Crane barge (Thor) lifting the second deck module (weighing 1750 tonnes) onto Graythorp I in the Forties oil field in the North Sea in 1974. (Photograph by courtesy of The British Petroleum Co. Ltd.)

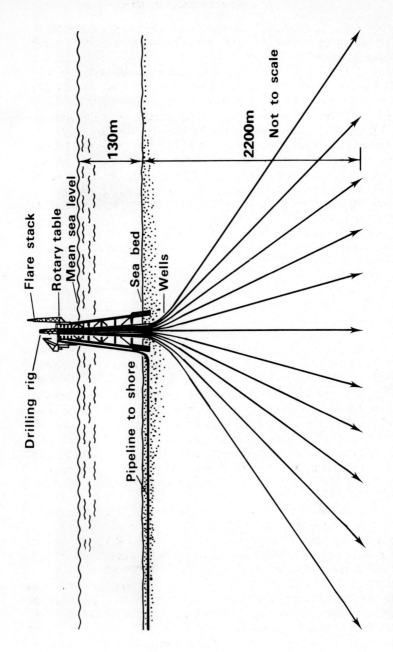

Fig. 7.11 Diagram illustrating drainage of wide area of North Sea oil field by wells fanning out at depth from a central platform.

damage can be compared with the forces which caused them. The design of future platforms can then be improved. Regular inspection of structures by divers using visual, photographic, and non-destructive testing methods allows engineers to measure corrosion and cracks. Finally, when an old platform is no longer needed it can be cut up and taken ashore for inspection to see what elements have deteriorated most dangerously, and which have survived best. Many lessons can be learnt for subsequent design.

A larger public is concerned about the environmental impact of offshore exploration and production. Mariners are concerned with obstacles to navigation; fishermen are concerned with sea floor obstacles (and, it may be remarked, the oil industry is equally concerned about the potential damage which may be caused by the impact of heavy trawls on sea floor equipment), while coastal towns and holiday makers are concerned about the possibility of major oil spillage.

Different phases of the operation provide both different kinds of hazards and different degrees of risk. The initial geophysical work involves the firing of explosive charges in the sea; in general this produces no problems, although with very large charges fish in the immediate vicinity may be stunned or killed, but the effect is very localized. It is only in the vicinity of special fisheries or spawning grounds — such as lobster beds — that restriction is called for.

Drilling is a relatively innocuous activity in terms of pollution, no worse than that of any normal seagoing vessel. The first few days involve drilling into the sea bed without casing, temporarily muddying the immediate area, but once casing is set the drilling fluid circulating in the well should be completely under control. Drilling fluid itself is a water or oil based mud with a range of chemical additives which may include phenols, and at some stage this is discharged as waste into the sea. The quantities however have not so far been sufficient to create perceptible problems, since the degree of dilution in sea water is very large. The most noticeable effect of a drilling rig in fact is to lead to a concentration of fish around the sea floor wellhead, either for shelter or for food waste thrown overboard.

It is at the drilling stage however that the danger of a blow-out is most critical and, correspondingly, the stage at which the most extensive precautions are taken. Firstly, the specific gravity of drilling fluid is increased so that the hydrostatic pressure at the bottom of the bore hole is greater than that of any expected fluid reservoir. Stand-by tanks of drilling mud are available in case of partial loss of the mud into the formation. Secondly, steel casing is cemented in the well at progressively deeper stages as drilling proceeds to provide a high pressure protection to overlying formations and to provide an anchor for the controlling valve systems. Well head valves are attached to this casing with a capacity considerably greater than any likely reservoir pressure, and these include blow-out preventer valves by which the well can be closed automatically, either

Fig. 7.12 Pipelaying barge (Viking Piper) laying the submarine pipeline from the Ninian oil field to Shetland in 1976. (Photograph by courtesy of The British Petroleum Co. Ltd.)

free or with the drill pipe in position. Provided that this equipment is properly installed and maintained in accordance with established regulations (including being tested every few days) any contingency should be controllable, and the risk to the environment negligible.

Nevertheless, the standard of housekeeping does occasionally lapse with disastrous results. (A recent offshore blow-out in the Middle East occurred when the blow-out preventers had been disconnected although drilling was in progress.) In that event a well may seal itself off after a short time; alternatively relief wells are drilled from rigs located as close as is practicable to intersect the flowing formation near the foot of the wild well, and to flood the formation with heavy drilling fluid, producing shut-off at depth.

Production testing is a less risky operation since by that stage underground pressure conditions are well understood. It is the usual practice to burn-off the resulting flow of gas or oil, which in the early stages is contaminated with drilling mud. This operation produces large quantities of smoke but should not result in any marine pollution. In the following phase of steady production oil is entirely contained within the system, but excess associated gas may be burnt off in special flares — formerly a universal and permanent feature of oil production, nowadays a temporary stage pending arrangements for gas collection and utilization, and one which is under strict government control.

The pipeline connections to shore should be buried and safe from danger from anchors or trawling. The burial operation is not always successful, and bottom currents may exhume a well buried line. Both conditions have occurred in the North Sea, and lines have been hit by heavy trawls. The thick steel pipe wall is, however, sufficiently strong to survive most accidents, and despite these contingencies breakage of a sea floor line is extremely infrequent. Regular inspection of pipelines by diver or submersible vehicle is a normal procedure and this should lead to discovery and repair of damage before critical deterioration occurs.

The final stage, the shipping of crude oil from the source area to the refinery, remains by far the most subject to disaster, as exemplified by the wrecking of the Torrey Canyon and Amoco Cadiz in recent years. Neither of these disasters had the excuse of bad weather, emphasizing that the human element remains dominant in producing large scale pollution. Present methods of dealing with spillage — particularly the organization of the operations — leave much to be desired, although the introduction of non-toxic detergents has limited the detrimental effect of bringing oil into solution. Damage to coastal fauna and flora and amenities remains extensive; at sea, fish mortality is on a minor scale but surface feeding birds are most at risk and may be killed in thousands. Rigid control of tanker routeing and much better government and inter-government coordination of procedures for high speed action on spillage are both high priorities.

8 Oil and Gas from Deep Waters

Since the deep oceans cover something like three-quarters of the surface of the globe their potential for hydrocarbon generation, accumulation and production is a matter of serious concern. It must be said immediately that their prospectiveness is much lower than that of the continental areas and their margins. The deep oceans provide no great hope for solving a future energy problem, but they are not entirely without prospects and hence justify consideration of the intriguing problems which they present. The word ocean will be used throughout this chapter to mean the ocean basin beyond the continental shelf.

Hydrocarbon shows are not unknown in deep water, and two notable cases have been found by the Deep Sea Drilling Project. One was rock impregnated with light oil on the Sigsbee Knolls in 370 m water depth, on a feature associated with diapiric salt intrusions in the sediments of the western Gulf of Mexico. This is a case of continuation of continental shelf type of sediments and structure into deep water. Another was natural gas encountered in boreholes through the Miocene in the Antarctic waters of the Ross Sea, in quantity sufficient to provide a drilling hazard.

The ocean floors, however, present a range of strongly contrasting environments — the continental slopes, continental rises, abyssal plains and mid-ocean ridges (Fig. 2.4, p.7). Within this range, the continental margins may be either tensional ('aseismic') as around most of the Atlantic and Indian Oceans, or compressional as on the subduction margins of most of the Pacific. Each of these regimes presents different degrees of prospectiveness.

As stated above the basic requirements for petroleum production include source rocks (organic-rich sediments), reservoir beds and cap rocks. To these descriminants must be added a suitable thermal regime — temperatures sufficiently high to generate hydrocarbons from the organic source material, but not so high as to have led to its destruction by cracking into the constituent elements. The thermal environment of the sediments is one of the most critical factors in relation to the prospects of the oceans.

Temperatures increase downwards in the earth at a fairly regular rate — about 11°C per km on the continents and 14°C per km beneath the ocean floors. In continental areas conditions suitable for conversion of the source compounds into fluid hydrocarbons are commonly found at depths of 2000–6000 m. The circulating bottom waters of the deep oceans are, however, cold, and the top of the sediments kept down to a temperature not far from 0°C. Thus downward increase in earth temperature has to be rated from that point; the depth of water above the bottom is irrelevant (except in relation to pressure conditions) and makes no positive contribution to the temperature regime. Although the temperature gradient beneath the ocean floor is believed to be generally higher than that on land, one of the resulting requirements is still a sedimentary column thick enough for the required conditions to be reached within it, independently of water depth.

One result of the low temperatures of the deep ocean floor is the accumulation of methane hydrates. Methane gas is generated in fresh sediments ranging from those of village ponds to the deep seas, and under the conditions of low temperature and high pressure in the ocean floor environment, accumulates as liquid hydrates, which are relatively immobile (as compared with gas), so do not easily escape and are widely present. Production of hydrates in liquid form may well be impractical and would certainly be uneconomic, but as the temperature increases downwards during sedimentary accumulation buried hydrates will become unstable and free methane released. If, under very exceptional circumstances, this methane was concentrated in significant quantities in the sediments it could provide extractable gas accumulations independently of the temperature parameters which are believed to control conventional hydrocarbon generation. No such case has yet been reported, however, and this is at present an entirely hypothetical concept.

Reverting to the more normal type of hydrocarbon occurrence, it must be emphasized that the different types of ocean floor provide widely different prospects. The mid-ocean ridges, occupying one third of the total oceanic area, can be dismissed as totally non-prospective: they consist of basalt flows underlain by packed sheets of igneous dykes, with only thin sediments in small rifts and tilt-block basins resting on the igneous basement. The dykes are believed to continue through the crust, so that there is no possibility of prospective deep sediments beneath a basalt cover.

The abyssal plains provide a different problem. Drilling in deep water has usually penetrated Tertiary and sometimes Cretaceous rocks which are largely deep-water clays and oozes with little potential for hydrocarbon generation and accumulation. They rest on basalt, the latter commonly described as 'basement'. However, there is no certainty that this igneous floor represents the beginning of the sedimentary history of the abyssal plains. It has in fact been indicated by seismic survey of the African coasts that the ocean floor basalt is equivalent to, and perhaps in continuity with, the Karroo basalt of inland Africa (Beck and Lehner, 1974), which implies that older sediments are at least locally, perhaps extensively, present beneath the basalt floor. Their petroleum potential must however be highly speculative: hydrocarbon generation and trapping may have occurred in suitable sediments, but the later history of burial by (probably) major thicknesses of basalt in an environment characterized by high temperature gradients is likely to have cracked liquid hydrocarbons to gas, and perhaps further metamorphosed gas to its constituent elements.

The ocean floor, however, includes not only the mid-ocean ridges and the abyssal plains, but also the so-called 'microcontinents' which are blocks essentially of continental rock (sial) commonly with a sedimentary cover, and with a possibility of peripheral sedimentary fans. These are believed to be fragments of the originally continuous continents which have been isolated in the course of plate movement. Examples include the Seychelles, Madagascar and the Kerguelen Plateau in the southern Indian Ocean, the Walvis Ridge off South Africa and the Orphan Knoll in the north Atlantic. Since any such fragments will be derived from continental edges they are likely to have some of the attributes of continental shelves, which are among the more prospective marine areas. Exploration of some of these areas is already in progress, but is still at an early stage. Thickness of sediments sufficient to have generated hydrocarbons is a major concern, since in most cases their isolation implies remoteness from source of detritus.

There is an intermediate category, consisting of sections of continental edge which are only partially separated from the present continents, by water which, although deep, is considerably less than full oceanic depth. Examples include the Blake Plateau adjoining the Florida coast and the Rockall Plateau off north-western Britain, both of which lie within the continental margin defined by the 5000 m isobath. A still closer degree of continuity is represented by areas of continental shelf which may have been faulted down into deep water without any element of lateral separation; these include the deep Exmouth Plateau off the north-western Australian coast and the Porcupine Bank area west of Ireland. The latter illustrates some of the problems of interpretation. It was formerly assumed, on the basis of bathymetry, to have been produced as a split by late lateral separation of a block from the European continent, but as the geology becomes known it is evident that the intervening Porcupine Bight is due to relatively ancient vertical subsidence, probably pre-dating the opening of the North Atlantic. The geological history of subsidence is critical in determining the prospectiveness of such areas; apart from adequate sedimentary thickness it is of course necessary to have effective source rocks and reservoir beds. These are most commonly accumulated in shallow water, dating from a period before major subsidence of the continental edge took place. Deep water pelagic sediments are much less likely to be economically productive, but can provide an effective cover on older porous sediments.

The more widespread type of continental margin shows the tripartite arrangement of shelf, slope and rise described above (Fig. 2.4). The continental rises, which are sedimentary fans with extremely gentle gradients, may have an effective thickness of sediments near the continents, but for the most part are likely to be made up of deep water sediments with little or no development of reservoir beds, and perhaps also an absence of source rocks.

The building of a continental slope is a complicated sedimentological process. In part the bulk of sediments is increased by building outwards from the continent – by deposition of sediments on the face of the slope. Sediments deposited in this way will be of shallow-water type at the top, grading into deep-water types down-slope. The whole mass is, however, subject to slow subsidence, so that sediments initially deposited in fairly shallow water, or on the outer continental slope, may come to be sited many thousands of metres deeper. Sediments deep in the continental slope are not, in consequence, to be automatically discounted as lacking in hydrocarbon reservoir beds of shallow water origin.

There is a further characteristic of continental slopes – a tendency for poorly consolidated sediment to slide down the face of the slope, producing landslips of hundreds of cubic kilometres of sediments, features which are on the whole decidedly detrimental to development of deep water resources. There are however some cases, as off the Niger delta, where gravity sliding has produced at the foot of the slope large thurst anticlinal structures which may themselves be prospective (Lehner and de Ruiter, 1976).

Additionally, a distinction has to be made between the prospects of Atlantic-type (passive) and Pacific-type (subduction) continental margins. Most existing offshore oilfields are on the continental shelves of Atlantic type. These developed by an initial phase of tensional faulting, with deposition of coarse sediments and evaporites in marginal rift valleys, followed by building-out of a massive sedimentary wedge as the ocean subsequently opened by lateral shift of the facing continental masses. Shallow water sediments, perhaps including porous carbonates deposited in the early stages of marine transgression, were deeply buried as they subsided beneath the accumulating sedimentary wedge. Such sediments (with reservoir and cap rock potential) are likely to occur beneath at least part of the continental slopes, although reservoir characteristics may be less good than those nearer the sources of the sediments. Development of deep water sands deposited by turbidity currents is an alternative, and reservoirs of this kind may be productive (as in the North Sea Forties oilfield) although they are more often disappointing (as in New Zealand). The problem of reservoirs in this part of the continental shelves has been discussed in more detail by Beck and Lehner (1974) and by Warman (1978).

The coasts of the Pacific type are generally less prospective in deep water than on the shelves. The geology of subduction zones still presents major unsolved problems, but it is generally held that the ocean floor is continuously underthrusting the continental plate or marginal sea, that oceanic sediments are being carried down to great depth and destroyed, and that older sediments are likely to be deeply buried and metamorphosed. The Pacific sedimentary wedges commonly include a major proportion of volcanogenic sediments which have poor porosity and permeability characteristics. Turbidity currents played a large part in sedimentary deposition and the sands they deposited are, in general, of low prospectivity – a generalization to which the California basin, with granitic sources of sediments, provides a striking exception.

Making guesses in the almost complete absence of accurate information on sediments on the deep ocean margins. Warman (1978) has suggested that the Atlantic and Indian Oceans might each yield 7.5 x 10^9 billion barrels of oil and the equivalent thermal value of gas, the deep Pacific margins perhaps much less, but in any area economic fields will be very difficult to find. These figures, which could be much too low or much too high, together, for all the oceans, total much less than (for example) the known reserves of Iran or recent discoveries in Mexico.

Location of favourable sites for exploration, using seismic survey, is perhaps the easiest stage of exploration in deep water, although the present procedures with both energy source and recording carried out near surface may have to be modified in favour of the sea floor operations for greater accuracy. Work on such equipment is in progress.

Drilling holes in the sea bed by the procedures of the Deep Sea Drilling Project is relatively simple, but these make no provision for carrying rock cuttings to surface or, more importantly, for controlling flow of fluid from formations penetrated in boreholes. Exploration for hydrocarbons, and their subsequent production, require much more sophisticated methods than these.

The basic requirement is a riser, to connect the borehole top with surface, several kilometres above. This must be strong enough to support its own weight (under tension) and that of a load of drilling mud which may be 5–20% heavier in gravity than sea water. Control valves have to be located on the sea floor at the bottom of the riser with operating lines (hydraulic or electric) to surface. An elaborate casing programme below the sea floor is required to ensure that the hydrostatic load of drilling mud does not exceed the strength of the rocks penetrated. Establishment of production will require sea bottom well control, including a remotely operated valve system, sea floor gas separators and flow lines to a (probably distant) coast or tanker loading point (Fig. 8.1). Some of the problems – those of deep sea floor well heads and remotely operated valve systems – are under active development already. Experience gained in the North Sea and other deep shelves has established general principles for controlled drilling on the ocean floor, which have been summarized by van Eek (1978).

The conclusion from this brief review is the pessimistic one that the deep oceans are overall of low prospectiveness for hydrocarbons; that

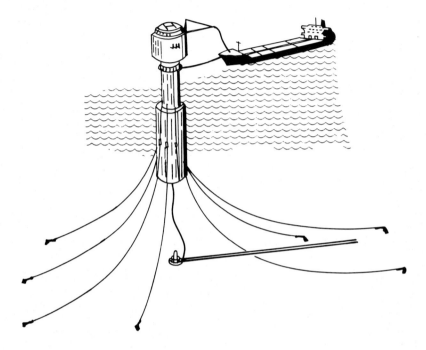

Fig. 8.1 Single-point moored loading system for water too deep for a conventional platform.

much of the area has nothing to offer, much of the rest is of very doubtful value, and apart from the unknown prospects of microcontinents only the continental slopes may be able to make any important contribution to world petroleum resources. Even these operations will be under circumstances of difficulty in location, exploration and production far greater than those of the continental shelves. Nevertheless, industry is proceeding into progressively deeper water, with production in 120–180 m water depths now conventional in the North Sea and with production exceptionally beneath more than 300 m of water, with controlled drilling operations in more than twice this water depth and riser-equipped drilling ships now capable of use in 1 800 m of water. Given encouragement from deep discovery, industry will certainly find the means for developing hydrocarbon production in water of oceanic depths.

9 Coal beneath the Sea

It has been emphasized above that continental shelves are the flooded edges of continents, similarly showing sedimentary basins separated by areas of older rocks. Just as hydrocarbon-prone basins continue offshore, coal bearing basins on land also have their offshore extensions. In some areas — as on the Durham coast (Fig. 9.1) and the Whitehaven area of Cumberland — mining has continued beneath the sea from coastal collieries, an arrangement analogous to the development of the Long Beach oilfield in California by directional drilling from the coast.

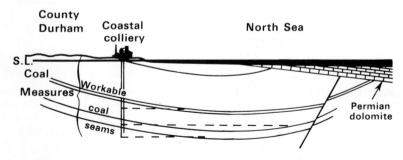

Fig. 9.1 Section showing subsea coal mining off the Durham coast.

Continental shelves are, however, characteristically areas of subsidence and continued sedimentation, vertical movement which has resulted in accumulation of barren overburden which may be measured in thousands of metres. Thus, whereas Coal Measures may be easily accessible in marginal areas, they are commonly too deeply-buried for conventional working over most continental shelves.

Britain and its surrounding seas provide examples of the situation and the problem of extracting the buried energy. Continuation of the East Midland—Yorkshire coalfield eastwards to the North Sea coast has been proved by deep boring, but at the mouth of the Humber the top of the Coal Measures is already at 1700m depth and the better seams are 100—400m deeper — well beyond the depth of economic working under present conditions.

Exploration of the southern North Sea has shown that the Coal Measures are continuous from the British coast eastwards to the Netherlands, but the depths are greater still. Associated with the increase in depth is a rise in coal rank, that is in carbon content, product of a very mild degree of metamorphism due to the increased load and the depression of the coal bearing beds to the hotter, deeper levels. Most of the 150—200 boreholes drilled for gas in the southern North Sea only touched coal measures beneath the gas bearing Permian sandstones, but some were drilled on for considerable distances, providing a thousand metres of coal measures with thick seams (Fig. 9.2).

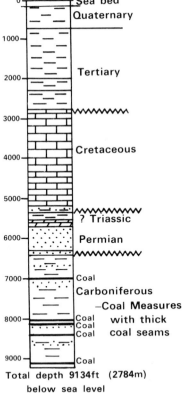

GULF borehole 15/10-1

Total depth 9134ft (2784m) below sea level

Fig. 9.2 Coal Measure sequence beneath the southern North Sea, showing the presence of deep seams which might be worked by underground gasification in the future. (Depths in feet.)

Another area of interest so far only indicated by geophysical (seismic) survey, is the English Channel and the Celtic Sea, where quite deep simple sedimentary basins have been located beneath the Permian, but above, the strongly folded Carboniferous rocks well known in Devonshire, must be filled with later Coal Measures (analogous to the Saar and other basins in France and western Europe), likely, but not yet known, to contain major coal reserves at depths of 2000–5000m.

Not all sedimentary basins found by geophysical work are coal bearing: a well defined basinal depression beneath the Thames estuary proved on drilling to be filled with barren Old Red Sandstone, any Carboniferous rocks having been eroded away.

How can the energy locked up in this vast coal measure basin be recovered? It is possible that there are more marginal areas accessible to conventional mining from land, for example the edge of the North Sea basin adjoining the north Norfolk coast. But the greater part of coal resources on continental shelves will be at depths of three thousand metres or more. Some extension of operations to greater depth will no doubt be made possible by development of fully automated mining equipment – 'mechanical moles' of various kinds which can locate, cut, load and follow a seam without direct human aid. Such equipment is well on the way to practicality (at high cost), and no doubt inclined shafts could be sunk to great depths by automatic equipment if necessary, despite the problems of great rock pressures and fluid ingress.

A more hopeful procedure would be in recovering the energy of the coal without mining it. Removal of coal by solution is one possibility, although the economics look far from favourable. Underground gasification, by which coal seams are converted into carbon monoxide and methane by partial burning *in situ*, is the most promising way of recovering the millions of tons of coal beneath the coastal seas, although despite years of experimentation practical methods for gasification *in situ* still remain to be developed.

To the extent that offshore coal is worked from coastal collieries, environmental problems will be familiar ones. With modern methods of pit head design and waste disposal the direct physical impact on the environment can be minimized: the main problems may well be social ones concerned with importation, housing and transport of thousands of miners in an alien milieu.

When methods are developed for extraction of the energy of offshore coalfields by underground gasification, it may be expected that the surface equipment will be comparable to a large oil-production platform, with compressors, gas scrubbers etc., serving numerous wells radiating over several square kilometres below the seabed. Again the environmental problems will be the familiar ones related to navigation, the laying of submarine pipelines and the connection with the national gas grid on land. But underground gasification as a means of working coal seams still lies years ahead.

10 Regional Review

It has been emphasized in the foregoing chapters that both the distribution
of the mineral resources of the world's seas and the scope for their individual exploitation are strongly dependent on environmental considerations.
It was circumstances of geology, past climates, oceanic circulation and
sedimentation which controlled the location of the wide range of materials
concerned, and it is circumstances of weather, sea conditions, water
depths, sea floor stability and accessibility which decide whether they are
economically recoverable in competition with more conventional mineral
sources.

This linkage between the geological environment and practical exploitation can be illustrated by a brief review of resources of the world's continental margins, omitting the deep oceans which involve the unique and
largely separate problem of the manganese nodules. It should be added
that this review is unlikely to be fully accurate or complete, since offshore and land minerals are not usually classed separately in official trade
figures, but it will give a general idea of the stage of development of
different minerals in different continents. Figure 10.1 gives a summary
of the range of mineral distribution.

North-western Europe

The dominant factor controlling the type and incidence of economic
minerals around the shores of north-western Europe has been the recent
glaciation, when most of the region was covered by thick ice which left
behind a layer of boulder clay blanketing many low lying areas of bed-
rock, and from which flowed vast areas of outwash sand and gravel.
The positive gain from this late history is the enormous availability of
sand and gravel for construction purposes. Up to the present this need
has largely been supplied from land areas but growth of population and
rising agricultural values make the offshore resources increasingly important
and attractive. They are worked in the North Sea, off the Thames, off
Scotland, in the Baltic and around Spitzbergen.

A negative factor of the glacial over-running of the surface has been
the removal of residual mineral deposits of the kind which are found in
tropical regions. There is one exception — placer tin has been worked in

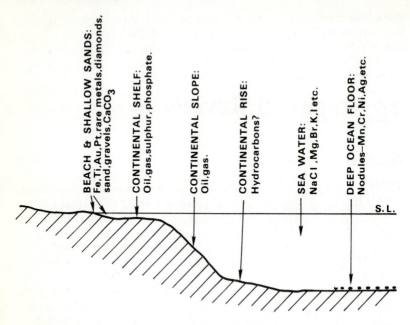

Fig. 10.1 Summary diagram showing the range of mineral distribution in the world's seas and oceans.

submarine alluvial gravels off Cornwall, and will no doubt receive attention again when the price structure makes it attractive.

Deeper minerals were of course unaffected by the glacial episode. Coal workings continue undersea from the coalfields of Cumbria, Northumberland and north-west Germany, and the development of gas and oil in the North Sea is a major economic success of modern times.

Eastern North America

Eastern Canada, like Europe, was overrun by glaciers during the Ice Age, but further south the coastal belt was free from this complication. In the middle coastal states the metamorphic rocks of the Appalachians outcrop one or two hundred kilometres in from the coast, and these have given rise to sands which are a source of titanium, zircon and monazite. The states near the Gulf of Mexico lack both coastal sands and calcium carbonate, and the offshore carbonate sands of the Bahamas Banks fill an important gap in local needs.

The search for deeper resources has given mixed results. Texas and Louisiana have enormous hydrocarbon reserves both onshore and offshore, but exploration results in Florida, in the Bahamas, and along the eastern seaboard have been negative or disappointing. There is now news of gas discoveries in the Baltimore Canyon off Maryland, and a gas field has been established one or two hundred kilometres off the Canadian coast near Sable Island, but the East Coast does not provide a province comparable with the Gulf of Mexico or the North Sea. This may be a relief to conservationists who have opposed development offshore in the region, but it represents a serious limitation in the long-term energy resources of the region.

Western North America

The western side of the Americas is marked by the Cordilleran belt, characterized by volcanic rocks, major granite intrusions and extensive mineralization. It is thus in line with the land geology that the beaches and offshore sands contain gold, platinum, titanium, zircon, monazite and other minerals in detrital form, continuing the landward alluvium which supported the historic gold rushes of California and the Yukon. The dissolved minerals of the sea — salt, magnesium, bromine etc. — have been extracted on a large scale for West Coast chemical industry, as great in value as the gold, with the advantage of being a renewable resource. California's oilfields are past their peak, although some discoveries continue to be made, but despite sophisticated exploration the remainder of the west coast of U.S.A. and Canada (south of Alaska) has not yielded more hydrocarbon reserves. Off California development is taking place beneath more than 500 metres of water, and in this sector further offshore discoveries may be made.

Alaska is a particularly favoured section of the west coast with beach sands rich in the minerals derived from the land area, including the gold sands of Nome, with barytes worked offshore, with offshore oil and gas fields in the Cook Inlet and a prospect of others off the north coast, reflecting a particularly complex geological history.

South America

Geologically South America is comparable in structure to the North, with a cordillera along the Pacific coast and broad plains which include outcrops of ancient and granitic rocks extending to the Atlantic. A comparable distribution of economic minerals might be expected, and both monazite and titanium minerals are recorded on the east coast. There seems however to have been little or no commercial development so far. An area of shelf floor east of Argentina has a potential for phosphorite exploitation, but it is an area notorious for extreme weather conditions and operations will be difficult.

Except for gas fields off Trinidad in the Caribbean and small oil-bearing areas in deep seated rifts off Brazil, South America appears to lack offshore hydrocarbon bearing basins, but it is a region still inadequately explored.

Australasia

Australia consists of a basement shield, largely older Palaeozoic and Precambrian rocks with a localized veneer of sediments on land, with deep marginal troughs round the coast. The older rocks of the continent are the source of the detrital beach sands which have been extensively worked along the east and south-west coasts, providing sources of titanium, zircon and monazite. Tin bearing sands have also been worked off New South Wales, and garnet sands are mined as an abrasive.

The marginal sedimentary troughs have proved to contain hydrocarbon bearing rocks — notably oilfields in the Bass Straits and large gas fields, still to be developed, on the broad north-western continental shelf. Sedimentary phosphates are known on the continental shelf surface, but they have not yet been found to be economic in competition with land sources.

New Zealand contrasts with the Australian continent in being a land area formed by late folding, and in having a major volcanic element in its constitution. Black iron sands occur derived from the volcanics, but have not been worked. The late sedimentary basins contain hydrocarbon bearing sediments, and a large gas field has been developed off the Taranaki coast of the North Island.

East Indies

The central East Indies from Thailand and Malaysia to Borneo include large granite intrusions which are a major source of the world's tin. Most of this is mined from placer deposits, originally on land, more recently from the sub-sea continuation of the fluviatile sands dating from periods of lowered sea level. The large western islands, Sumatra and Java, are geologically new mountain ranges flanking late Tertiary sedimentary basins, and the continuation of these basins in offshore Java, Borneo and the Philippines has proved to be the site of further oil and gas fields.

The Indian Subcontinent

Peninsular India and Ceylon provide another case of an ancient continental nucleus which has contributed resources of ore bearing sands to its coastal regions. The coast of Ceylon is particularly rich in monazite, rutile and titanium minerals, as well as gem bearing sands.

The Indian peninsula is flanked by sedimentary basins which have a potential for hydrocarbon exploration, but results have so far been limited. An offshore field has been found off the mouth of the Indus (on the structure known as the 'Bombay High'), but much of the Bay of Bengal is blanketed by a vast cone of sediment from the Ganges river system and no discoveries have yet been made.

Africa

Coastal Africa has been less intensely investigated than many parts of the world, and its full potential has still to be realized. It is another case of a block of ancient rocks, rich in minerals, which has given rise to local developments of ore bearing sands along its coasts. The occurrence of diamonds off Southwest Africa is an outstanding case, although their working has been suspended since more easily won on-shore sources can currently supply the market. Titanium (monazite and zircon) sands have been worked on the south-east coast and on the coast of Madagascar, and there are deposits of zircon and titanium known also on the west African coast. Iron sands are worked off Sierra Leone. Phosphorite deposits in the recent sediments off the Cape of Good Hope and in the Gulf of Guinea may be important in the future.

Again it is peripheral basins of Mesozoic and later sediments which have provided hydrocarbon reserves, initially onshore but now also off the coasts of Nigeria and Gabon and in the Red Sea. The distribution of hydrocarbons is however very uneven, with only a small gas discovery off South Africa and no fields at all along the east coast of the continent. Offshore discoveries may be expected off the North African coasts in due course, but operations have not yet met with success.

In summary, the offshore activities round the continents at present are dominantly those of recovering detrital metal minerals from natural concentrates in coastal and shallow water sands, and the production of oil and gas from deep sedimentary basins. The occurrence of the former depends on the type of the hinterland and the environmental factors involved in sorting and concentrating heavy minerals. The latter are dependent on the existence of the sedimentary basins, and specifically on a combination of past geological conditions which led to generation, migration and entrapment, but not distribution, of hydrocarbons, both gaseous and liquid.

Extensive deposits of such low value material as phosphorite remain as a resource for the future, and it is only a few industrialized nations which have made any serious use of the vast tonnage of dissolved materials in the sea (other than salt) — of its magnesium, bromine and potash.

This brief review has thus tended to emphasize the limited extent to which the seas' mineral resources are still being used, and, given suitable economic conditions, the implications for wider exploration and development. Environmental problems and conservation must be balanced against the need for resources, such as comparing the damage done by the mining of beach sands with the need for the components of television screens. Decisions on such priorities are far from easy: there will be no facile solutions.

11 Recapitulation: Special Problems of Mineral Recovery in the Marine Environment

In extraction of dissolved minerals from the sea the constant movement of the water is an essential factor. For all other processes listed in previous chapters it is on balance an inconvenience, with only the removal and dilution to harmless levels of polluting waste to be counted in its favour. Land-based mineral industry can rely on firm foundations for equipment, and on uniform, predictable gravitational and kinetic forces affecting the machinery and material processed. At sea, floating equipment is not fixed in any direction; it is subject to a full range of spatial displacements and angular movements which operate continuously. Additionally ships and equipment must work in a corrosive atmosphere, often under very adverse conditions of temperature and humidity.

The basic problems of instability of the sea surface were encountered when oil rigs moved into water too deep for legged platforms. Except in calm weather the earlier moored drill-ships had to head to the seas, involving constant re-orientation and frequent interruption of drilling. Semi-submersible drilling barges provided much more stable drilling platforms, for their large flotation tanks at the foot of the legs are some 20 m below surface, and wave-induced movement of the platform is largely eliminated. They are still subject to the effects of long swells which are liable to lift the suspended drilling tools off the bottom of the hole or alternatively apply an undue amount of the weight on the suspended equipment; this factor is offset either by insertion of a splined telescopic section low down in the drill string or by hydraulically controlled compensators at surface, or both.

Changes in water depth due to waves, swell and tides affect dredging operations even more critically, since either a bucket ladder or a hydraulic lift is a relatively massive piece of equipment which must operate a fixed distance above the sea floor. The precise positioning is particularly serious for marine mineral placer deposits, which commonly occur at the contact of gravel with bedrock, and are at their richest in holes and depressions in the rock floor. Efficient dredging operations under these conditions requires a high degree of seamanship.

The immediate problem of water stability ceases in offshore oil drilling operations at the development stage, when a fixed platform may be planted in 200 m or more of water depth. The first problem in fact is stability of the sea floor as a foundation for the platform, since hydro-carbons occur in sedimentary basins which are often still receiving unconsolidated sediment, subject to irregular deposition and erosion under the influence of sea floor currents. Emplacement of a platform may lead to a local concentration of erosive activities and the undermining of a structure, especially at storm periods, as has happened for example in the North Sea. The sites for permanent platforms are carefully evaluated and they are heavily piled into the sea floor at a very early stage.

Whether fixed or floating, marine structures are subject to the continu-ous corrosive action of salt water and salt spray and also to continuous deformation under the stresses of waves and currents. One of the North Sea semi-submersible drilling rigs, found to have developed hair-line cracks in bracing members, was calculated to have been flexed by wave action some two million times in three years, resulting in fatigue fracture and requiring local reinforcement. Fixed platforms there and elsewhere have shown abnormal sensitivity to corrosion despite all conventional preventatives, and a further environment constraint was illustrated when it was found that a production platform designed for use in the Bay of Biscay could not be safely used in the northern North Sea because of the danger of steel embrittlement in the much colder waters off Scotland.

Nearly all metallic minerals are mined in low concentration, and a major part of their treatment has to be carried out before they are transported. Here again there is a major difference between the situation on land and sea; the separation processes are frequently gravity controlled and beneficiation normally requires a stable situation. Concentration over riffles, by centrifuge or by flotation are all liable to lose efficiency if taking place on a rolling ship at sea, and a great deal of research is taking place into modification of methods to cope with offshore circumstances. Roll, pitch, yaw and heave all exert stresses on the system which are quite large in relation to those of the processing operation; they have to be quantified and provided against in design.

An equally important but entirely different set of problems arises in relation to transport of sand, gravel or mineral concentrates from the source area to port. The material is wet when received and the concen-tration involves first washing and then removal of most of the water. But even a damp granular material, subject to the continuous vibration of propulsion and other machinery is liable to temporary liquefaction, or thixotropic behaviour, and as a cargo may become completely unstable to the extent of shifting and causing a vessel to capsize in a heavy sea. This has long been recognized as a hazard in relation to fine coal transport and sand and gravel dredgings; it will be true of mineral concentrates whether fine grained or granular; it will be a particular danger in the transport of the largely fine grained product of manganese nodule mining in the deep ocean. Use of transport ships with compartmented holds, as in an oil tanker, will appreciably increase handling costs but will probably be an essesntial feature of mineral concentrate shipment (Hughes, 1978).

Finally, the constraints of international law, or its absence, provide an overall limitation to a great deal of marine mineral exploitation. The Law of the Sea Conference has defined the oceans as the common heritage of mankind, but it is only a handful of developed nations which have the technical ability, economic and financial resources to deal with such problems as those of exploiting manganese nodules in the deep oceans, and a considerable degree of organizational stability is required to ensure the economic returns sufficient to justify the required major outlay in funds, material and manpower. Just as exploration in the North Sea was organizationally impractical until the riparian nations agreed on offshore boundaries and provided the stability of a licensing system, so a degree of guaranteed continuity in development rights is necessary for oceanic operations, whether for hydrocarbons or metals. The world may soon be desperate for the additional resources which the oceans can provide; we hope that the nations (and their politicians) will be able to produce a *modus operandi* for their utilization for the benefit of the world's population.

Glossary I: Geological Periods

Geological period names shown in their correct relative order

Era	Period	TIME RANGE (million years before present)
TERTIARY or CENOZOIC	PLEISTOCENE	1.8–0
	PLIOCENE	1.8–5
	MIOCENE	5–22.5
	OLIGOCENE	22.5–38
	EOCENE	38–55
	PALAEOCENE	55–65
MESOZOIC	CRETACEOUS	65–141
	JURASSIC	141–195
	TRIASSIC	195–230
PALAEOZOIC	PERMIAN	230–280
	CARBONIFEROUS	280–345
	DEVONIAN	345–395
	SILURIAN	395–435
	ORDOVICIAN	435–500
	CAMBRIAN	500–570
PROTEROZOIC		570–2600
ARCHAEOZOIC		More than 2600 million years before present

Glossary II: Technical Terms

Abyssal plain   The nearly flat floor of the very deep ocean

Algal bloom   Massive development of pelagic algae associated with upwelling currents

Anaerobic (adj.)   Formed in absence of oxygen e.g. sulphurous bottom sediments

Aragonite   Form of calcium carbonate, $CaCO_3$, often of biogenic origin

Barytes   Barium sulphate, $BaSO_4$, used in paint, paper making, etc.

Basalt   Basic igneous rock widely present on or beneath the ocean floor

Basement rocks   Term variously used for ocean floor basalts or for fundamental crystalline rocks underlying the sedimentary column

Bathymetry   Measurement of water depth, particularly in the ocean

Benthic   (Organisms) found on the sea and ocean bottom

Cap rock   Impervious layer (often of sulphate) overlying oil-bearing strata

Chromite   Oxide of iron and chromium, $FeCr_2O_4$, occurring primarily in ultrabasic rocks and occurring as a detrital mineral

Concretions   Segregations of carbonate or other minerals in sediments

Continental shelf   Submerged continental edge, usually ending at c. 200 m depth on the edge of the continental slope

Continental slope   Relatively steep slope (gradient c. 1 in 20) extending from the edge of the continental shelf to the continental rise at 3–500 m depth

Continental rise   Gentle slope (gradient c. 1 in 100) extending from the foot of the continental slope to the abyssal plain

Dolomite   Double carbonate of calcium and magnesium, $CaMg(CO_3)_2$, a major rock-forming mineral

Echogram   Plotted result of shallow acoustic survey

Fathom   Depth measurement, six feet, 1.83 metres

Fetch   Distance over which wind-sea interaction raises waves

Fraasch Process   Recovery of native sulphur by melting out with super-heated steam through boreholes

Glauconite   Green hydrous silicate of iron and potassium (variable composition) occurring disseminated in shallow water sediments

Gravity (gravimetric) survey   Measurement of variation in the earth's gravity field, indicative of density of underlying rocks

Guano   Rock phosphate produced by reaction of bird droppings with coral rock

Hydrates (of methane)   Solid combinations of methane and water existing at low temperatures under pressure

Igneous rock   Rock formerly in molten form (e.g. basalt, granite)

Ilmenite   Oxide of titanium and iron, FeO, TiO_2, commonly occurring as a detrital mineral

Kimberlite   Igneous vent material which is the primary source of diamonds

Lacustrine (adj.)   Pertaining to lakes

Laminaria   Seaweed extensively used for fertilizer and for recovery of minerals especially potassium, bromine and iodine

Lode   Vein of rock bearing metallic ore

Magnetometer survey   Survey method based on measurement of variations in the earth's magnetic field

M.I.A.S.   Marine Information and Advisory Service run by the Institute of Oceanographic Sciences, England

Mid-ocean ridges   Central mountainous belts in the deep ocean caused by a combination of tensional opening and extrusion of igneous rock

Magnetite   Magnetic iron oxide, Fe_3O_4, sometimes occurring in economic quantity in 'black sands'

Manganese nodules   Concretions dominantly of iron and manganese oxides, with minor but important proportions of nickel, silver, chromium etc. found on the deep ocean floor

Marginal plates   Structural units which together comprise the large plates around the Pacific

Monazite   Phosphate of the cerium metals (Ce, La, Yt) PO_4, usually with thoria, ThO_2 and silica, SiO_2, occurring as a detrital mineral

Pelagic   (Organisms) of the open ocean, usually near the surface

Phosphate   Natural calcium phosphate, usually with calcium carbonate, on oceanic islands (see guano) or disseminated on the sea floor

Phytoplankton   Microscopic plant material which forms the beginning of the massive food chain

Placer   Concentration of detrital minerals in a sand, forming an ore body

Rutile   Titanium dioxide, TiO_2, commonly occurring as a detrital mineral

Salt dome   An intrusion of massive rock salt into overlying strata, usually in the form of a steep-sided column

Sand waves   Parallel banks of sand in shallow water resembling sand dunes
Sea mount   Submarine mountain, usually flat topped, often of volcanic
  origin
Seismic survey   Large scale acoustic survey method recording reflected
  or refracted energy usually from an explosive source
Sial   Rock of acid composition making up the continental masses
Sima   Fundamental basic rock underlying oceans and continents
Sonic (survey)   Method of surveying using acoustic signals
Sparker survey   Shallow acoustic reflection survey using a spark-ignited
  explosion as energy source
Subduction   The process of underthrusting of oceanic crustal plates
  at continental edges, specifically on the Pacific ocean margins
Swell   Waves generated by wind beyond the limits of storm or continuing
  after wind has dropped

Terrigenous   Sediment of terrestrial origin
Tsunamis   Highly destructive major waves produced by earthquake action
Turbidity current   Gravity-induced subsurface flow of water heavily
  loaded with sediment, usually on continental slopes, causing submarine
  erosion

Upwelling   Large scale movement of deep ocean water to the surface

Volcanogenic   (adj.) Rock of volcanic origin

Zircon   Zirconium silicate, $ZrSiO_4$, commonly occurring as a detrital
  mineral

References

ARCHER, A.A. (1973). Economics of offshore exploration and production of solid minerals on the continental shelf. *Ocean Management,* **1**, 5–40.

ARCHER, A.A. (1976). Prospects for the exploitation of manganese nodules: the main technical, economic and legal problems. In GLASBY, J.P. and KATZ, H.R. (Eds). *UN Economic and Social Committee for Asia and the Pacific. Committee for Coordination of Joint Prospecting for Mineral Resources in South Pacific Offshore Areas (CCOP/SOPAC). Tech. Bull.* **2**, 21–38.

ARMSTRONG, E.F. and MIALL, L.M. (1946). *Raw Materials from the Sea.* Chemical Publ. Co., New York. 196pp.

BECK, R.H. and LEHNER, P. (1974). Oceans, new frontiers in exploration. *Am. Ass. Petrol. geol. Bull.,* **58**, 376–395.

HARRIS, S. (1978). Cashing in on the oceans – the new manganese Klondike. *The Listener,* 7th Sept. 1978, 295–296.

HUGHES, T.H. (1978). Effect of the environment on processing and handling materials at sea. In *Sea Floor Development: Moving into Deep Water.* Royal Society, London. pp. 161–177.

LEHNER, P. and de RUITER, P.A.D. (1976). Africa's Atlantic margin typified by string of basins. *Oil & Gas J.,* **74**, 252–266.

McILHENNY, W.F. (1975). Extraction of economic inorganic materials from sea water. In RILEY, J.P. and SKIRROW, G.(Eds.) *Chemical Oceanography,* Vol. **4**, 2nd Edition, pp. 155–218.

MEADOWS, D.H., MEADOWS, D.L., RANDERS, J. and BEHRENS III, W.W. (1972). *The Limits to Growth.* Earth Island Ltd., London.

MERO, J.L. (1965). *The Mineral Resources of the Sea.* Elsevier, Amsterdam. 312pp.

PRESCOTT, V. (1977). Mining the sea bed and the law of the sea. *Optima,* **26(4)**, 222–240.

SMALE-ADAMS, K.B. and JACKSON, G.O. (1978). Manganese nodule mining. In *Sea Floor Development: Moving into Deep Water.* Royal Society, London. pp. 125–133.

UNESCO (1971). *The Sea. Mineral Resources of the Sea.* 51st Session. Report of the Secretary General, E/4973 of 26.4.71.

van EEK, W.H. (1978). The challenge of producing oil and gas in deep water. In *Sea Floor Development: Moving into Deep Water.* Royal Society, London. pp. 113–124.

WARMAN, H.R. (1978). Hydrocarbon potential of deep water. In *Sea Floor Development: Moving into Deep Water.* Royal Society, London. pp. 33–42.

WEEKS, L.G. (1968). The gas, oil and sulphur potential of the sea. *Ocean Ind.*, **3(6)**, 43–51.

Index